Contaminated Communities

Contaminated Communities

The Social
and Psychological Impacts
of Residential Toxic Exposure

Michael R. Edelstein

Ramapo College of New Jersey

Westview Press
BOULDER & LONDON

Grateful acknowledgment is made for the permission to quote extensively from the following materials: Michael R. Edelstein and Abraham Wandersman, "Community Dynamics in Coping with Toxic Exposure," in Irwin Altman and Abraham Wandersman (eds.), *Neighborhood and Community Environments* (New York: Plenum Press, 1987) (copyright © 1987 by Plenum Press); Michael R. Edelstein, "Disabling Communities: The Impact of Regulatory Proceedings," *Journal of Environmental Systems* 16, No. 2(1986/ 1987), 87–110 (copyright © 1986 by JES); Michael R. Edelstein, "Toxic Exposure and the Inversion of the Home," *Journal of Architectural and Planning Research* 3(1986), 237–251 (copyright © 1986 by Elsevier Science Publishing Co., Inc.); and Michael R. Edelstein, "Social Impacts and Social Change: Some Initial Thoughts on the Emergence of a Toxic Victims Movement," *Impact Assessment Bulletin* 3, No. 3(1984/1985) (copyright © 1984 by International Association for Impact Assessment).

Copyright © 1988 by Westview Press, Inc.

Published in 1988 in the United States of America by Westview Press, Inc., 5500 Central Avenue, Boulder, Colorado 80301

Library of Congress Cataloging-in-Publication Data
Edelstein, Michael R.
    Contaminated communities: the social and psychological impacts of
residential toxic exposure/Michael R. Edelstein.
    p. cm.
Bibliography: p.
Includes index.
ISBN 0-8133-7447-2
ISBN 0-8133-7657-2 (if published as a paperback)
    1. Indoor air pollution—Social aspects. 2. Indoor air pollution—
Psychological aspects. 3. Environmental health—Social aspects.
4. Environmental health—Psychological aspects. 5. Pollution—
Environmental aspects. 6. Pollutants—Environmental aspects.
I. Title.
    [DNLM: 1. Environmental Pollutants—adverse effects.
2. Psychology, Social.   WA 671 E21c]
TD883.1.E347 1988
363.1'79—dc19
DNLM/DLC
for Library of Congress   87-31653
                                                                              CIP

Printed and bound in the United States of America

    The paper used in this publication meets the requirements of the American National
    Standard for Permanence of Paper for Printed Library Materials Z39.48-1984.

15 14 13 12 11 10 9 8 7

To the memory of my mother,
Evelyn S. Edelstein

Contents

1 Toxic Exposure: The Plague of Our Time   1

The Plague as a Metaphor for Toxic Exposure, 2
Contaminated Communities, 4
Defining a Contaminated Community, 5
Toxic Exposure as Disaster, 6
Theoretical Framework for the Book, 9
Summary and Outline of the Book, 14
Notes, 15

2 Legler: The Story of a Contaminated Community   17

The Legler Case Study, 17
The Period of Incubation, 18
Discovery and Announcement, 29
Disruption of Lifestyle—The Delivery of Water, 34
The Hookup of City Water, 38
Notes, 41

**3 Lifescape Change: Cognitive Adjustment
to Toxic Exposure**   43

Perceiving a Changed Status, 44
Perceptions of Health, 49
Environment, 54
Loss of Personal Control, 57
The Inversion of Home, 61
Loss of Trust, 70
Conclusion—The Lifescape Impacts
  of Toxic Exposure, 82

Notes, 82

Figures

Foreword

In the late 1970s, the Love Canal sounded a warning. Those who listened understood that a new type of disaster was emerging in our world in the form of an unanticipated price for the benefits derived from the post–World War II burgeoning of the chemical industry. Since that time, we have learned that the price goes well beyond money, for it also encompasses physical, psychological, and social distress. While the burden falls most heavily on individuals and families exposed to hazardous pollution, that price is being extracted as well from groups, neighborhoods, communities, and societies. It will continue to be paid long into the future, unless we collectively recognize that the human, social problems created by our technologies cannot be addressed simply in technocratic terms. Michael Edelstein is one of the people who heeded the warning of Love Canal. As he makes clear in this book, the problems must be addressed with an emphasis on the concerns, needs, interests, and rights of the people whom technology is supposed to benefit, in such fashion that our social fabric becomes stronger, or at least is not weakened.

In the late 1970s, when Edelstein, as a young professor, began to expand his academic studies of the environment to include the psychological and social effects of toxic pollution, he also expanded his own role to become an environmental activist. This book shows the results of his decade of work, which started at a time when the human problem of toxic pollution was scarcely recognized as an area of study for social scientists. In the ensuing years, the events at Seveso, Love Canal, Times Beach, and numerous other sites have drawn the attention of researchers who have produced a number of published pieces and papers read at conferences devoted to the issues. Meanwhile, a broad social movement has begun, which draws its members from communities threatened by toxic wastes and its inspiration from the example of the citizens' movement at Love Canal. Edelstein has drawn together literature from all these sources, and has coupled it with his own research and observations, to produce a coherent, soundly based picture of the broad, deep, and long-lasting effects of exposure to hazardous wastes, on every element of society.

Contaminated Communities is a solid addition to the literature of a new, growing field of study. The book is absorbing and enlightening because the author provides richly detailed pictures of affected people trying to cope with problems they had never anticipated having, problems that are so new that no one truly knows their dimensions, let alone what the solutions might be. It should be read by policy makers who will not only gain insight into the thinking of the affected people they are mandated to serve when toxic pollution strikes a community, but will also be able to reflect on the constraints and problems of their own roles in these cases. This book may help them avoid some of the pitfalls that have ensnared so many in the past decade, and have created alienation between some government officials and the people whose lives are massively disrupted by toxic pollution.

But *Contaminated Communities* should be read by a wide range of people. Edelstein makes us realize that in the broadest sense we are all policy makers, for we must all ask and try to answer the hard questions, we must all take responsibility to improve our society. It is books like this that will help us think about our choices, about the sort of physical and social world we want to maintain and create for ourselves, and leave as our legacy for generations to come.

Adeline Gordon Levine
State University of New York at Buffalo

Preface

When I think about residential toxic exposure, various scenes often flit through my mind. These are some of the poignant images that give meaning to all the words that will follow:

A five-year-old boy whose house abuts a toxic waste dump draws a picture of his home and its surroundings. In the center of the drawing is the building, crudely sketched as any five year old might depict his house. A "For Sale" sign dominates the front yard. In the backyard, a tall mound of red material is shown leaching residue under the house. Facing this mound, his back to the house, stands the boy. His face is shown with a massive frown.

A man and a boy pick large and beautiful flowers that they call "dioxin irises" from the backyard of an abandoned house on Love Canal.

A forlorn nurseryman gives me some plants so that they might have a chance to survive, away from his contaminated property.

A chainlink fence tacked with warning signs surrounds boarded-up homes. On the other side of the street, children ride tricycles along the sidewalk. A block away stands an abandoned elementary school.

A woman of late middle age describes the horrible sores on her breasts that she blames on fumes from the factory nearby. With a sense that all vestiges of privacy have been eliminated, she opens her blouse to show me.

*   *   *

There are many such images in my memory of the past nine years, during which I have devoted much of my work as an environmental psychologist to observing the social and psychological impacts of residential toxic exposure incidents. I have interviewed scores of toxic victims in their homes, most frequently so that I could serve as an expert witness in hearings or in lawsuits filed on their behalf. I have also acted as an environmental leader in my own community. Through it all, my work has taken me to a wide range of settings—rural, suburban, and urban—affected by a wide variety of contaminants. In some cases, exposure was well documented. In others, it was only suspected. In still others, concern over a potentially polluting facility was sufficient to generate surprising amounts of stress. From these experiences, I have become convinced that an understanding of the psychosocial impacts of residential exposure is central to developing effective responses to toxic disaster. It is an equally important consideration if we are to meet the challenge of waste disposal without creating the next generation of toxic hotspots. This volume is not intended to exhaust the topic but rather to encourage further study of the issues surrounding toxic exposure.

It is impossible for me to address this topic without being continually reminded of the deep hurt felt by many of those whom I have interviewed and encountered. I have never succeeded in keeping a "professional" distance by reducing people to mere "experimental subjects." Sharing the pain is part of the understanding. On more than a few occasions, I have been haunted by my experiences with toxic victims, not just during the interviews but throughout the entire period in which I analyzed and wrote about the case. Writing this book reawakened for me many of these poignant feelings. I have let my informants speak liberally on these pages (their comments appear in italics) so that the reader can also share as much as possible the intensity of this experience. At the same time, in an attempt to respect their privacy, I have not disclosed the true identity of any field informants.

The growing field of social impact assessment offers the potential to stretch beyond purely academic interests in order to use what is learned as a basis for mitigating and minimizing adverse impacts. Accordingly, as a researcher, I see myself also as an advocate, in the sense that full documentation of an issue is a step toward having that issue addressed. I hope that this volume proves to be an impetus for examining many of the institutional and social barriers that have in the past impeded efforts to address the needs of toxic victims.

It has been unfortunately common for industry and government to malign toxic victims as being overemotional "chemophobics." If there is any one realization that emerges from my work, it is that the stresses that accompany the discovery of and adjustment to exposure present so aberrant a context for victims, that to find little indication of stress might be a true sign of abnormality. As I once suggested to the lawyer representing a hazardous waste site, the best "cure" for the depression evident in the landfill's neighbor would be adequate compensation to let him get on with his life, not psychological counseling. Accordingly, I believe that the social and psychological impacts of toxic exposure must be "treated" broadly by society and that the social sciences can offer guidance.

Michael R. Edelstein
Mahwah, N.J.

Acknowledgments

I have many people to thank for help at many levels.

First, I owe a tremendous debt of gratitude to my informants. Your sharing and openness allowed me to gain insight into your experience without which I would have had little to say. In many cases, my opportunities to gain entry to the situations that I studied depended upon the lawyers who supported my research and their associates and paralegals. Steven Phillips, Ivan Rubin, Ellen Cookson, Maureen Barrett, and Michael Gordon deserve particular mention.

I have benefited from outstanding support from colleagues during this effort. My colleagues at Ramapo College of New Jersey granted me sabbatical and other research support. Colleagues deserving specific mention for their help at various junctures are Gordon Bear, Eric Karlin, William Makofske, Margaret Ottum, and Frank Sirianni. C. P. Wolf, of the Center for Social Impact Assessment, played a supportive and helpful role from the inception of the project. Maragret Gibbs, of Fairleigh Dickinson University, has been a close collaborator and friend throughout my work on toxic exposure.

A small group of colleagues played the particularly difficult but invaluable role of reading my various drafts and providing the feedback that I needed to bring the book to print. I greatly appreciated not only their positive support but also the times they chose honesty over kindness. Adeline Gordon Levine, of the State University of New York at Buffalo, went way beyond the call of friendship in her detailed readings of the manuscript. I also wish to thank Janet Fitchen, of Ithaca College; Lois Marie Gibbs, of the Clearinghouse for Hazardous Wastes; Marcia Mattheus, of the Goshen Area Resources Association; Abe Wandersman, of the University of South Carolina; and Deborah Kleese, of Empire State College. I also benefited from the help of a number of Westview Press staff, most notably my acquisition editors, Barbara Ellington and Krista Mueller, my project editor, Libby Barstow, and my copy editor, Ida May Norton.

The logistics of the book were accomplished largely on a Morrow Micro Decision computer. Additional support also came from Sherri Cox, Pat Dicker, and Pat Dominici, all of Ramapo College, and from

Leona Hoodes, of Orange Environment, Inc. Patricia Palmieri of Ramapo College assisted with the graphics.

Deborah Kleese served a particularly important role. Not only was she available as a colleague for soliciting opinions at almost any hour, but as my wife, her tolerance, support, and encouragement made the book possible. My son, Joel, likewise gave up a lot of his time with me to this project. He made up for some of it by sitting on my lap before the computer and wildly pressing keys at inopportune moments and otherwise engaging me in play therapy. I worked out many of my ideas while in the company of Walter, my Newfoundland, on long midnight walks.

Despite this lengthy list, there are others deserving my sincere thanks. To all who supported this endeavor, I hope that the result does credit to our efforts.

M.R.E.

Contaminated Communities

1

Toxic Exposure:
The Plague of Our Time

In *A Journal of the Plague Year,* Daniel Defoe chronicled the transformation of everyday life that accompanied the Great Plague of London
of 1665, which killed some 100,000 people. As news of the plague
abroad reached them, Londoners nervously began to watch the "bills
of mortality," the weekly reports of number and cause of deaths in each
parish, which served as indicators of the plague's appearance. Many
fled the plague, leaving the city comparatively empty. The wealthiest
were most able to leave. Others, in the trades or without resources,
were generally forced to remain, only to face unemployment resulting
from the disruption of London's economy. Many who left the city perished
on the road from hunger and want of lodging.

A plague culture emerged within London. Fear of strangers became
the norm. Books foretold the ruin of the city; on the streets there appeared
those prophesying destruction. Others set up practices to treat those
infected, some offering "Charms, philtres, exorcisms, amulets . . . "
(Defoe, 1960, p. 40). In their alarm, people confessed crimes long
concealed. Many believed that heaven had sent the distemper. At first
people mourned the dead, but then the expectation that their own death
would shortly follow hardened them. "A kind of sadness and horror
. . . sat upon the countenances even of the common people. Death was
before their eyes, and everybody began to think of their graves, not of
mirth and diversions" (p. 37).

Parents, not knowing they were diseased, inadvertently infected their
children. The Lord Mayor began shutting up sick people in their houses
on July 1, 1665. Watchmen were set at their doors. Because entire families
were quarantined with the sick persons, all were doomed to perish.

No one knew how the infection spread—giving fuel for various
superstitions. People came to define any suspicious occurrence or symp-
tom in light of the plague. Medical experts were of little help in clarifying

the situation, arguing over whether open fires might help to control the plague's spread and, if so, what type of trees might best be burned. Not only was the daily life of Londoners dramatically affected by the plague, but their entire way of seeing and comprehending the world was altered.

The Plague as a Metaphor for Toxic Exposure

An awareness of society's responsibility for public health was born during the time of plagues. With the growth of commercial centers and the industrial revolution, urban density was recognized as a factor in health problems. Technologies were developed to provide sanitation, control of vectors (carriers of disease), and other protection against bacteriological disease (see Dalton, 1973). But industrialization also bequeathed a less beneficial legacy—environmental pollution. For example, in *Hard Times*, Dickens described rivers "that ran purple with ill-smelling dye." Thus, ironically, even as advances were made in public health, environmental pollution created even more problems.

Although today we are beginning to recognize the implications of pollution, our concept of public health until recently has retained its exclusive focus upon problems of sanitation. We have been ill-prepared to confront a new type of threat: exposure to toxic substances found in our air, water, and food. Toxic exposure is fomenting yet another revolution in the conception of public health; in this sense, it is the plague of our time.[1]

Recognition of the Issue

Not until well after World War II was public attention drawn to toxic exposure. Dramatic air pollution disasters involving smog in Donora, Pennsylvania, in 1948, London in 1952, and New York City in 1953, 1963, and 1966 had left thousands of people dead or injured, giving impetus to concerns over air pollution that jelled in the 1960s (Eckholm, 1982). Not until the 1970s, did the United States begin to address toxic contamination resulting from the common use of synthetic chemicals that also affected water and food.

Scientists following the lead of Rachel Carson (1962) have brought to light many of the environmental impacts of toxic exposure. In this book, I address other impacts of toxic exposure that, although little recognized heretofore, are perhaps of equal importance. These are the social and psychological effects caused both by residential toxic exposure and by the social response to exposure.

The Extent of Contamination

The dimensions of toxic exposure are staggering, reflecting our dependence upon an increasingly synthetic world.[2] Some 70,000 chemicals are in regular use in the United States and another 1,000 are added every year. This includes 1 billion pounds of pesticides, herbicides, and fungicides used in the United States every day. Beyond toxic exposure due to the manufacture, transportation, storage, and use of these materials, this country generates between 255 million and 275 million metric tons of hazardous waste annually, of which as much as 90 percent is improperly disposed of. Such figures routinely exclude the military, historically one of the worst polluters. In 1984 alone, more than 530,000 tons of hazardous waste were produced at 333 U.S. military installations.

The results of this societal chemical-dependence are shocking, if not surprising. The Office of Technology Assessment estimates that there are some 600,000 contaminated sites in the country, of which 888 sites have been designated or proposed by the Environmental Protection Agency for priority cleanup under the Superfund program, with another 19,000 sites under review. In addition, there are another 400,000 municipal landfills, more than 100,000 liquid waste impoundments, millions of septic tanks, hundreds of thousands of deep-well injection sites, and some 300,000 leaking underground gasoline storage tanks threatening groundwater. Hundreds of planned and operating garbage incinerators are generating new concerns about air pollution and highly toxic ash residues. And this list merely scratches the surface of causes for residential toxic exposure. A complete chronicle of residential toxic exposure would also include the countless sites where wastes have been illegally or inadvertently dumped; the existence of naturally occurring toxins such as radon gas; exposure to radioactive wastes; toxic substances in building materials, home heating sources, foods, and household items; and the widespread use of pesticides in agriculture, forestry, lawn care, termite control, utility right-of-way maintenance, and in routine home applications.

What is perplexing about so numbing a list of statistics is that it took so long to recognize a phenomenon of such magnitude. Excellent journalistic accounts of toxic incidents during the 1970s produced global awareness of such disparate tragedies as mercury poisoning in Minimata, Japan; the dioxin release in Seveso, Italy; the contamination of food by mercury in Iraq; and Agent Orange poisoning in Southeast Asia. But the awareness of Americans was particularly aroused by numerous domestic toxic incidents, the most dramatic being discovery of a major toxic waste dump beneath a residential community in the Love Canal section of Niagara Falls, New York. It bore witness to the new plague—toxic exposure.

Contaminated Communities

Although detailed case study material appears in later chapters, it may be helpful here to provide thumbnail sketches of a number of contaminated communities.

Contamination as a Widespread Event

As a heavily publicized event, Love Canal perhaps symbolized best the realization that toxic exposure can destroy a neighborhood[3]. An uncompleted canal, dug late in the last century to create hydroelectricity for industry, Love Canal was used by Hooker Chemical and others as a dumpsite for chemical wastes beginning in the 1940s. In the 1950s, a residential neighborhood and school were developed along the canal. When widespread chemical contamination was discovered in 1978 by federal officials studying pesticide pollution of the Great Lakes, the implications of residents being exposed to the dumped chemicals began to unfold. Concerned over existing and future health problems, residents organized to represent their interests and to advocate government action. Using public demonstrations and political pressure, as well as their own scientific studies, residents were eventually able to achieve government-sponsored relocation from Love Canal. As of 1987, Love Canal is largely abandoned, despite controversial plans to resettle it.

Even as the events at Love Canal captured media attention, residents in hundreds of communities across the United States were confronted by similar circumstances[4]. In northern New Jersey, residents of the Relocated Bayway section of Elizabeth, as well as their neighbors in Staten Island, New York, faced extreme airborne exposures in April 1980 when between 50,000 and 60,000 barrels of chemical wastes exploded and burned at the Chemical Control facility, only one of numerous polluters in that neighborhood. In the early 1980s, a continuing pollution problem was uncovered with the discovery that radioactive mill tailings, used for housing fill in a number of Essex County, New Jersey, communities, posed a threat to residents due to elevated levels of radon gas and gamma radiation. In the summer of 1983, extremely high amounts of the potent chemical dioxin were discovered in the ethnically diverse Ironbound section of Newark, New Jersey. These problems were not confined to the urban areas; in 1980, bicyclists riding along a rural road in Warwick, New York, just over the New Jersey border, discovered 22 seeping barrels of organic chemicals illegally dumped on an alfalfa field near a cluster of homes relying on well water for consumption.

Meanwhile, in Woburn, Massachusetts, a high incidence of childhood leukemia was explained when it was discovered that industrial solvents had contaminated local groundwater. A General Motors dumpsite leached extremely dangerous compounds (PCBs, dioxins, and furans) onto the Akwesasne Indian reservation in upstate New York; an upturn in birth defects among the Mohawk children born there was thus explained. Furthermore, the 9,000 Mohawk faced having to eschew a traditional lifestyle dependent upon hunting and fishing upon sacred lands because those lands had been tainted by pollution and the resident game poisoned. In Tuscaloosa, Alabama, a new garbage incinerator, built in the early 1980s to solve the area's garbage disposal problems, quickly generated complaints about strong odors and pollutants that made residents ill. The plant also generated fly ash found to contain significant levels of the heavy metals cadmium and lead and was inoperational so often that it caused a financial crisis. Throughout the 1970s, a waste hauler named Russell Bliss, using dioxin-contaminated oil to treat roads and horse arenas throughout Missouri, laid the basis for a widespread dioxin crisis across that state that caused illness in people, killed horses, and resulted in the federal buyout of the town of Times Beach.

The list of contaminated communities is virtually endless. In the summer of 1982, Minnesota officials announced extensive groundwater contamination by the chemical trichloroethylene in two suburbs of Minneapolis, forcing residents to find new sources for drinking water. Meanwhile, residents of parts of the strikingly beautiful Skagit River Valley in northern Washington State, already beset by unemployment due to a downturn in the forest industry, faced further problems because the pesticide EDB, sprayed on strawberry crops in the valley, leached into their groundwater. In Aurora, Colorado, residents became alarmed at the effects and risks associated with a long-operational 2,600-acre landfill found to contain cyanide, solvents, heavy metals, asbestos, pesticides, and organic compounds. In 1978, after a decade of concern, California neighbors of the Stringfellow Acid Pits organized when officials, fearing a break in the dam holding back 32 million gallons of toxic chemicals, pumped 1 million gallons into the community. Across the state of Louisiana, a string of pollution incidents began in 1978, incidents that have since been linked with data on sharply increasing cancer rates. And residents of Midland, Michigan, home to a large Dow Chemical complex that has pumped up to 300,000 gallons a day of some 500 chemicals into the ground for thirty years, have shown various symptoms of dioxin poisoning.

Defining a Contaminated Community

Scholarly attention to the psychosocial impacts of toxic exposure in communities such as these dates from the efforts of Levine (1982) to document the Love Canal disaster. Since then, a small but growing number of social scientists have begun to develop a conceptual understanding of toxic exposure.

For purposes of analysis, the events at Love Canal, as at hundreds of other locations, suggest the need to draw boundaries around an area identified as being polluted. Accordingly, I will use the term "contaminated community" to refer to any residential area located within the identified boundaries for a known exposure to some form of pollution. Whether or not residents share a similar political, geographic, or social environment, the discovery of a toxic threat provides a basis for a new and shared identity that effectively defines a community of interest among those residing within the boundaries of contamination.

Toxic Exposure as Disaster

As an ecological event, community contamination involves the immediate or gradual deterioration of the relationship between humans and the ecosystem (Couch and Kroll-Smith, 1985). As with naturally occurring disasters, such as floods and hurricanes, victims of toxic exposure experience stress because their way of life is disrupted and society cannot readily restore what was lost (see Barton, 1969). But to a much greater extent than with most natural disasters, the "facts" of toxic disaster are often unclear, making the "perception" of the disaster central to its subsequent effects. Threats to health and safety, social relationships, and the prevailing worldview are likely to enhance the perceived extent of the disaster (Couch and Kroll-Smith, 1985). The resulting stress may affect the ability of victims to return to work or social activities. Trauma associated with disaster affects the family and community as well as the individual (Erikson, 1976; Janis, 1971). Various studies have suggested that human-caused disasters result in greater, longer-lasting, and different kinds of stresses than those associated with natural disaster.[5]

Two distinguishing characteristics of toxic disaster particularly influence its perception. First, toxic disaster is technological in origin. As such, it implies that the technological control over nature that enables us to enjoy a high standard of living has failed. The resulting "loss of control" contrasts with our "lack of control" over natural disasters (Baum et al., 1983). This reaction to technological disaster is evident in the increasing public concern over technological dependence (Kushnir, 1982; see also Milbrath, 1984).

The second characteristic is illustrated by contrasting the perceptions of victims of the 1977 Johnstown, Pennsylvania, flood with those of residents of the Love Canal neighborhood. While the first group tended to see their disaster as "an act of God" from which they needed to rebuild, residents I interviewed at Love Canal in 1979 regarded their disaster as stemming from corporate greed and government corruption. The human-caused origin of the toxic disaster invited attributions of responsibility that greatly colored perceptions of the event (see Wolfenstein, 1957). Similar outcomes are found with other human-caused disasters. For example, survivors of the Buffalo Creek flood in West Virginia, where coal slag dumped into the creek collapsed to cause the disaster, did not blame God for bringing the rain; they blamed the coal company for damming the creek (Erikson, 1976). As these examples suggest, when victims realize that a disaster is human-caused, they are likely to develop feelings of distrust and anger toward the perceived agents.[6]

The Stages of Toxic Disaster

Toxic disaster occurs in stages that roughly correspond to those commonly used to describe natural disaster.[7] These stages describe most toxic incidents, although there may be exceptions because of such factors as the type of contaminant, the mode of contamination, population characteristics, the acuteness of the exposure, certainty about the consequences, and quality and quantity of assistance.

Pre-Disaster Stages: Origin and Incubation. The originating circumstances for a case of toxic exposure may vary, but there is usually an "incubation" stage, when the community is unaware that the disaster is developing. Therefore, there are no preparations or premonitions (see Baum et al., 1983).

There are a number of reasons for the failure to predict and recognize an incubating toxic disaster. For the eventual victims, the disaster is likely to be a novel experience, rather than a recurrent one; the unfamiliar circumstances do not trigger suspicion. And because such disasters are not supposed to happen, it is assumed that they won't. Furthermore, pollution is frequently barely detectable, not only in occurrence but also in consequence. The threat may be invisible, and any resulting damage may be hard to relate to the pollution (Miller, 1984; Baum et al., 1983). The subtlety of the clues makes them easy to discount. Finally, it is generally assumed that government is watching out for us. Yet agencies we look to for protection are often not vigilant in monitoring environmental hazards or do not share information. Together, these factors reflect the complexity of ecological and technological problems and the uncertainties involved in toxic disasters that make the risks hard to define (de Boer, 1986; Miller, 1984).

Disaster Stages: Discovery, Acceptance, Community Action. Because a toxic disaster strikes gradually and incrementally without people's knowledge, detection of the problem, warning of potential victims, and their perception of threat may occur long after the disaster has struck. Accordingly, much of the initial psychosocial impact is due to the announcement of the disaster and the sharing of information about it. Beliefs formed at this point may persist even in the face of new evidence at a later time.

Experts specially trained to measure and detect toxic substances are usually involved in the process of discovery and announcement (Miller, 1984). Victims become dependent upon others to help them understand the situation and to help create solutions. However, because the effects of human-caused-disaster may not be readily visible, not only is the occurrence subject to differing interpretations, but key decisions (e.g., regarding testing, protective measures, and remediation) may be based on these interpretations (Miller, 1984; Couch and Kroll-Smith, 1985). Therefore, consensus about the cause, course, and possible outcomes of the crisis is less likely than with natural disaster. Furthermore, because there may be no visible damage (Baum et al., 1983; Couch and Kroll-Smith, 1985), each family is forced to make its own determination of the significance of contamination (Fowlkes and Miller, 1982). The lack of shared beliefs about what has happened opens the way for conflict within the community and between the community and potential helpers (Levine, 1982; Couch and Kroll-Smith, 1985).

Ironically, because they are not anticipated, human-caused disasters are routinely experienced as sudden even when they actually develop gradually (Baum et al., 1983). Barton (1969) notes that while a gradually occurring disaster allows the existing social system to make adjustments, the sudden-onset disaster requires a new process for reducing chaos.

Despite the difficulty of accurately identifying the affected area given the pervasiveness of pollution (Miller, 1984), at the earliest opportunity government officials commonly draw boundaries around the area they believe to be affected. Therefore, the boundaries of the disaster are socially clear, even if the criteria for drawing them are scientifically fuzzy. The community defined by the pollution boundaries becomes isolated from its surroundings, not by destruction, but by stigma. Yet, the collective isolation of victims sharing common concerns frequently contributes to cohesiveness around a new community organization. Because the community is trapped intact in the situation rather than destroyed, residents tend to donate a great deal of energy to this organization. That activism parallels changes in lifestyle and beliefs that occur at the individual and family levels.

Post-Disaster Stages: Mitigation and Lasting Impacts. While natural disaster is often of brief duration (Barton, 1969), human-caused disasters such as toxic exposure may be chronic and indefinite (Baum et al., 1983;

Edelstein, 1982). A site may be contaminated so that it will remain unsafe for generations due to the persistence of the toxic hazard; individual effects may also cross generations (Hohenemser et al., 1983). A sense of finality is elusive for the toxic victim (Edelstein, 1982), in part because toxic disasters lack a "low point" from which things would be expected to improve (Baum et al., 1983).

Because it is not clear what damage has occurred to property or finances, or what long-term health effects may develop, it is difficult to inventory losses. Basic needs such as water may be allocated on an emergency or even permanent basis, but rarely is rescue quickly forthcoming. Helping professionals may not define the events as a disaster to which they must respond. Bureaucrats may have vested interests in not helping in a definitive way. Friends and relatives may not know how to help. As a result, the affected community is left in a vacuum to fend for itself.

Remedy may involve mere mitigation, such as the provision of a new water source or a law suit that produces some form of compensation. But toxic disaster is hard to completely remedy because its impacts are impossible to measure. Much as with the prediction of contamination moving underground, it is difficult to gauge what the genetic implications of toxic exposure will be—neither of these outcomes can be seen, and the effects may not surface for quite some time. As a result, recovery to a "post-disaster equilibrium" is difficult if not impossible.

Theoretical Framework for the Book

This volume seeks to identify the major social and psychological impacts stemming from residential toxic exposure and to examine their significance. The theoretical framework for analysis is based on four postulates:

- The social and psychological impacts of toxic exposure involve complex interactions among various levels of society. They also differ across time and environmental context.
- These impacts not only affect how victims behave but how they perceive and comprehend their lives, in both the short and long term.
- Toxic exposure incidents are stressful, forcing victims to adopt some form of coping response.
- Contamination is inherently stigmatizing and arouses anticipatory fears.

Each of these postulates will be briefly discussed.

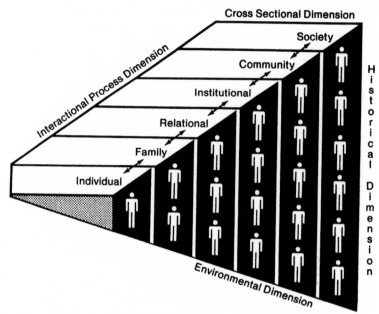

Figure 1.1 The Social Process Model

Postulate One: The Impacts of Toxic Exposure
Occur at Interacting Levels of Social Process

Toxic exposure is likely to entail social and psychological impacts both from the exposure itself and from the social context of the disaster. Accordingly, responding effectively to the disaster requires attention to psychological as well as physical parameters. Engineers and physical scientists know how to approach the latter, but a conceptual framework is needed within which social, psychological, and cultural effects of toxic exposure can be comprehended. The model adapted for this purpose is termed the "social process" model. As depicted in Figure 1.1, the social process model recognizes that any complex event, such as toxic exposure, has impacts at various levels of society both simultaneously and over time. Accordingly, any occurrence affects individuals, families and other relational groups (kinships and friendships), more formal associations (institutions and organizations), the community, and the society. The first function of Figure 1.1 is to show the inherent mutual interaction of these different levels of process.

Each of these contexts can in turn be examined in light of two additional dimensions depicted in Figure 1.1, history and environment. The historical dimension reflects the recognition that groups, organizations, and communities are influenced by their histories much as is the individual. A final dimension is environment, both the physical and social. While the relevance of the physical environment to toxic exposure is self-evident, the social environment is equally germane. Thus, the response of an institution to a toxic disaster reflects not merely the influences of other levels of process or the institution's own history, but also its broader social environment. For example, a given regulatory agency is probably only one of many agencies fighting for a common pool of resources. The availability of those resources is, in turn, a key factor in the agency's response to a toxic disaster.

Postulate Two: Toxic Exposure Affects
Both Action and Cognition

At each level of social process, one can distinguish a pattern of functioning that reflects both a normal set of behaviors, the "lifestyle," and a normal framework for understanding the environment, which I term the "lifescape."

Lifestyle refers to people's way of living, including their pattern of activities and the relationships and props needed to sustain these activities. Lifestyle embodies the core assumptions of a society as reflected in the pursuit of more personal goals. The achievement of these goals and social expectations reflects what is loosely called the "quality of life," encompassing such factors as economic security, family life, personal strengths, friendships, and the attractiveness of the physical environment (Campbell, 1981).

Beyond our routine activities, toxic exposure may disrupt the lifescape, a term that refers to our fundamental understandings about what to expect from the world around us.[8] The lifescape reflects not only our own unique interpretive framework, but also the shared social and personal paradigms used for understanding the world. We are rarely aware of our lifescape until something disrupts our lifestyle or in some other way disconfirms our operating assumptions. The crisis resulting from this disconfirmation is often resolved through denial; we do not easily abandon an existing explanatory framework. But if a new explanation is identified that better accounts for the anomalous situation, we may change our core understandings.

Beyond disrupting personal paradigms, disconfirmation of the lifescape may challenge core assumptions of the overall society. For example, toxic exposure directly assails several fundamental social beliefs: that humans have dominion over nature; that personal control over one's destiny is possible; that technology and science are forces of progress only; that risks necessary for the good life are acceptable; that people get what they deserve; that experts know best; that the marketplace is self-regulating; that one's home is one's castle; that people have the right to do what they wish on their own property; and that government exists to help. It is not easy to discard such beliefs, at least unless one has previously come to doubt the dominant paradigm. Some people who suffer toxic exposure and who have a strong belief in the social paradigm may even deny, rationalize, or minimize their exposure.

Postulate Three: Toxic Exposure
Is Inherently Stressful

For the individual, the physiological, psychological, and social costs exacted by everyday life are commonly referred to as stress. Toxic exposure may cause new strains in victims' lives, intensify existing strains, or give new meaning to old problems (see Pearlin et al., 1981). It may be blamed for such major life crises as the death of a loved one, marital discord, ill health of family members, and financial difficulties. None of the key spheres of life activity—home, work, friendships—is free from disruption due to exposure.

Stress is not solely the property of the individual, however. It occurs at every level of social process, affecting family, relational, institutional, community, and societal dynamics as well (see Monat and Lazarus, 1977). For example, at the institutional level of analysis, one can observe that response to a toxic exposure case affects individuals within an agency, the public image of the agency, its institutional standing as an arm of government, its success at meeting organizational goals, and its use of available resources. Given that stress exists at all levels of process, it may be meaningful to speak of a "system's" reponse to stress, rather than merely an individual's response.

Stress begins with a threatening occurrence that disrupts the balance or steady state achieved by the system (Lumsden, 1975). Stress-inducing threat can take three forms—noxious physical stressors, the realization that some event either may cause disruption or has already caused some harm or loss, and challenges, varying from simple problems to complex dilemmas (see Baum et al., 1981; Lazarus, 1966). Toxic exposure often involves all three types of threat simultaneously. Noxious physical conditions and fears over future consequences disrupt victims who are already overloaded with novel, complex, and ambiguous information that they need to understand and use in making decisions.

Once a threat is recognized, it is appraised by the system to ascertain the potential implications, their severity (Lazarus, 1964; Janis and Mann, 1977), the system's susceptibility, and corrective measures that can overcome effects (Beck and Frankel, 1981; Rogers, 1975). A decision is reached as to whether the threat can be discounted and ignored or whether further attention is demanded. In the latter case, a survey of available alternatives is undertaken. Because appraisal is a subjective analysis, varied interpretations of a given event are likely. While some systems take a threat seriously, others may engage in defensive avoidance (Janis and Mann, 1977). Factors affecting appraisal include the beliefs of the appraisers, their knowledge of the threat, the threat's visibility (e.g., amount and kind of publicity), and the significance of the threat (Edelstein, 1984).

Once a threat is appraised as legitimate, the affected system must consider the alternatives. In the absence of easy solutions or choices, the system may be forced to continue its existing practices. Thus, for example, most toxic victims remain in their homes despite concerns over risk of continued exposure. Some victims will cope with this imbalance (e.g., "I'm at risk, but I'm staying") by engaging in a form of denial. Other residents, however, regard themselves as stranded in a hazardous situation and experience an increased sense of threat because they lack alternatives. This situation results in what Janis and Mann (and most regulatory officials) refer to as "panic." Finally, the system may develop a vigilant decision-making approach in which members actively survey and evaluate alternatives, searching for the most desirable outcomes (Janis and Mann, 1977).

Coping responses vary, depending on whether they are focused at helping the affected system control the disruption or at influencing the source of the threat.[9] Accommodation to stress frequently occurs either because people adjust their behavior or beliefs or through adaptation because their sensitivity to environmental stimuli changes (Sonnenfeld, 1966). Examples of adjustment include the use of background sounds (music or air conditioners) to filter traffic noise (Edelstein et al., 1975) and the personalization of pollution detection equipment in living rooms by giving the equipment names such as "Old Harry" and "Darlene."[10] Examples of adaptation include the Love Canal resident who reacted to being told that the house smelled just like the chemical plants by commenting, "We got so used to living in the house, so the odor could have been there all the time" (Fowlkes and Miller, 1982, p. 82). Similarly, Evans and Jacobs (1981) found that the reaction of new residents to air pollution corresponded to what they had grown accustomed to in their previous location. Such individual accommodations of stress have parallels in relational groups, institutions, and communities as well. However, while they assist in coping with stress, accommodations do not necessarily reduce the perceived severity of environmental problems (Preston et al., 1983; Wohlwill, 1966).

Postulate Four: Exposure Is Inherently
Stigmatizing and Arouses Anticipatory Fear

Stigma is a consequence of being "contaminated" (Edelstein, 1987; 1984). Stigma always involves a victim identified by an observer as marked (deviant, flawed, limited, spoiled, or generally undesirable). When the mark is noticed, it changes in a negative and discrediting way how the observer sees the victim, whose identity is now spoiled (Jones et al., 1984; Goffman, 1963). And because we tend to assume that people deserve what happens to them, stigma readily invites a tendency to blame the victim (Ryan, 1971). In pollution cases, stigma routinely accompanies the announcement of contamination and the identification of its boundaries. The stigma can apply to a variety of targets, including affected residents, objects, places, animals, and products.

Jones and his colleagues (1984) suggest a number of criteria used by observers for evaluating a stigma that are applicable to toxic exposure. Accordingly, one can evaluate how disruptive an exposure event is, whether its existence is concealable, whether it affects aesthetic qualities, whether the victim is in any way responsible for its occurrence, what its prognosis is, and the degree of peril it portends. Furthermore, according to these authors, "the essence of stigma is fear" (1984, p. 65). Once contaminated, many exposure victims view themselves differently, in part because they fear dreaded health impacts, such as cancers, threats to unborn children, and cross-generational genetic effects. Victims also discover that others see them differently as well. Their homes and neighborhood are downgraded by observers who exhibit "anticipatory fears" about the place.

Anticipatory fears are perceptions of threat associated with future outcomes that are connected causally to current happenings (see Janis, 1971). The initial lack of vigilance that results in toxic exposure reflects an absence of anticipatory fear. But once contamination or a potential for contamination is recognized, much greater vigilance can be expected. After people mentally rehearse an anticipated danger, they may take protective actions. Drawing on the experience of others, people may even learn to rehearse disasters that they have not personally experienced and, therefore, to anticipate threatening events not as yet set in motion. Opposition to stigmatized facilities, such as waste disposal sites, is partially a consequence of anticipatory fears associated with them.

Summary and Outline of the Book

The impacts of toxic exposure need to be analyzed across the levels of social process. They involve effects on both lifestyle and lifescape and can be explained in light of a stress process involving threats, appraisal, and coping efforts, which include accommodation as well as more assertive and successful actions. Toxic exposure also inherently invites stigma and anticipatory fears. These four postulates form the framework for analysis in what follows.

In Chapter 2 of this book, I use a detailed case study to describe the lifestyle impacts of toxic exposure. In Chapter 3, five lifescape changes resulting from toxic exposure are reported. In Chapter 4, I discuss the cumulative family impacts, highlighting individual and relational impacts as well as effects on children. Chapter 5 describes how toxic victims are disabled, or rendered powerless, by the institutional context of an exposure incident. Chapter 6 explains how citizens regain power through grass roots organizations. Chapter 7 examines reactions to potentially threatening and stigmatized facilities. Chapter 8 draws conclusions about the social implications of toxic exposure and the role of social science in addressing it.

Notes

1. When I first drafted this chapter, toxic exposure was *the* plague of our time. However, AIDS has emerged as another competing plague over this period. Despite its idiosyncracies, AIDS fits the model of the biological plague, whereas toxic exposure introduces the technological plague that brings biological effects from physical changes to the environment wrought by human hands.

2. While there is much variation in estimates of contamination, all point toward a dramatically widespread phenomenon. These numbers are drawn largely from an excellent unpublished paper by Margaret Ottum written in 1983 as well as from Ridley, 1987. Other sources were Office of Technology Assessment, 1983; Robertson, 1983; Hewitt, 1981; and Toth, 1981.

3. See Shaw and Milbrath, 1983; Fowlkes and Miller, 1982; L. Gibbs, 1982a; Levine, 1982; Brown, 1980.

4. These examples were drawn from various sources, including the following: CCHW, "Tuscaloosa's Turkey," *Everyone's Backyard*, 5, No. 2(1987), 1, 3, 6; Culver, Alicia, "More than a Hole in the Ground," *Exposure*, 40(July/August, 1984), 1; Culver, Alicia, "Louisiana: Learning to Spell Environment," *Exposure*, 30(May, 1983), 5; Durso-Hughes, Katherine and Jim Lewis, "Penny Newman," *Exposure*, 15(February, 1982), 4–8; Egan, Terence, "Across the Street from Danger," *Sunday Record (Middletown, New York)* (April 3, 1983), 5; Gordon, Ben, "The Canary in the Coalmine," *Toxics in Your Community Newsletter* (September, 1986), 2–10; Kolata, Gina, "Missouri's Costly Dioxin Lesson," *Science*, 219(January 28, 1983), 367–368; Parisi, Albert, "Newark Farmer's Market: Dioxin Imperils Future," *New York Times* (Sunday, June 12, 1983), New Jersey Section, 2; Sheppard, Nathaniel, Jr., "Toxic Chemicals in Drinking Water Disrupt Life in 2 Suburbs of Minneapolis," *New York Times* (Dec. 26, 1982). The best survey of contaminated communities is Brown (1980).

5. See Baum et al., 1983; Edelstein, 1982; L. Gibbs, 1982a; Levine, 1982.

6. See also Couch and Kroll-Smith, 1985; Baum et al., 1983; Wilkenson, 1983; and Edelstein, 1982.

7. See Barton, 1969; Miller, 1964; Wallace, 1957. For another approach to describing the stages of toxic disaster, see Finsterbusch, 1987.

8. There has been surprisingly little conceptualization of lifestyle. Lifescape as a concept is derived from the work of Kuhn (1962) on paradigms of science, as well as from the efforts of several writers to broaden the paradigm concept into the study of dominant social frames of reference (see Harmon, 1976; Pirages, 1978; Milbrath, 1984; Devall and Sessions, 1985). In psychology, see Janoff-Bulman and Frieze, 1983.

9. See Hatcher, 1982; Evans and Jacobs, 1981; Lazarus, 1966.

10. Kroll-Smith and Couch, private communication regarding Centralia, Pennsylvania.

2

Legler: The Story of a Contaminated Community

I was born in the steel town of Johnstown, Pennsylvania, where the billowing smokestacks and flashes of fire from the blast furnaces were viewed as signs of prosperity, not sources of pollution. Generations had come and gone around the mills. To us, they belonged in the landscape, much as did the mountains, the valley, and the rivers. I doubt that newcomers and tourists can appreciate this architecture of industry in the same way.

Contrast this with the attitude of longtime residents of a mill town studied by Evans and Jacobs (1981) who viewed their mill negatively. These old-timers identified with the town as it had been before the mill; for them the mill was an intruder that violated their expectations for the quality of the local environment. Newcomers were comparatively more accepting. Their initial image of the town included the industry and its impacts. In deciding to move there, they had already accommodated to the conditions that existed. And some may even have moved from a more polluted area and so viewed their new setting as comparatively pristine (see Wohlwill and Kohn, 1973).

The Legler Case Study

As these examples suggest, the baseline expectations residents hold about their homes are an important indicator of how they will view subsequent events. Therefore, this examination of the impacts of toxic exposure begins with a comparison of victims' lives before and after an exposure event, the first serving as a baseline for analyzing the second. A case study of groundwater contamination in the Legler section of Jackson, New Jersey, serves as the basis for the comparison. The case derives from a detailed analysis that I prepared[1] in 1981 for the law firm that represented ninety-six Legler families belonging to the Concerned

Citizens Committee in a legal action against Jackson Township and others (Edelstein, 1982). The study specifically sought to identify the social and psychological impacts on these families resulting from the pollution of their well water by the adjacent municipal landfill.

I carefully chose twenty-five families whose views would be representative of all ninety-six families. At least one adult member of each of the selected families participated in a lengthy, intensive interview during which I tried to reconstruct the way Legler was viewed at the time the respondent had moved there, as well as to capture the person's perceptions of other key junctures in the chronology of the Legler incident.

All of the selected families shared two important similarities: They were middle-class, and they were homeowners. But they differed in significant ways as well. I chose the families to be interviewed according to four of these differences.[2] The first three factors related to residents' possible feelings of vulnerability to toxic exposure. The fourth reflected their response to the situation.

Age was the first factor. Legler was dominated by married couples under forty years of age. Not only did most already have children at home, but they were young enough to produce more. And given their age, they were themselves susceptible to slowly developing diseases such as cancer. As a result, I expected that their concern over health effects, both for themselves and their chidren, would be quite different from that felt by older residents.[3] In picking the sample, therefore, I tried to assure a mix of ages.

The second factor in choosing the sample was differences in length of residence; the third was direction and distance of the home from the pollution source. As two possible indicators of the degree of exposure, these factors also reflected vulnerability to harm from the incident. Activism was the final factor; the sample included all members of the executive committee of the Legler Concerned Citizens Committee as well as people comparatively much less involved.

In asking questions and in analyzing the data, I sought to understand how residents interpreted events, rather than to test hypotheses quantitatively. The number of times a given response occurred was, therefore, less important than the depth of each response and the contribution of the response to elucidating the overall picture. Using this qualitative approach, I was able to reconstruct respondents' residential expectations and to chart many of the impacts of the pollution incident.

The Period of Incubation

During the period before pollution was discovered, early residents formed their residential expectations for Legler. Later, they were forced to adjust to the development and operation of the municipal landfill, the source of contamination.

Residential Expectations in Legler

Legler is distinctive in its basic similarity to many other residential areas lacking its notoriety. Located to the southwest of New York City, Legler is a suburb overlaid onto an older rural landscape in a way that parallels thousands of other developments of the past few decades. In every way, the area symbolizes the search for the "American dream."

Achieving the "American Dream." Located at the edge of the sandy Pine Barrens in a remote section of Jackson Township, Legler is surrounded by open spaces. Atlantic Ocean beaches are nearby. The section reflects the intensive development that characterized Jackson and all of Ocean County during the fifteen years preceding the discovery of contamination. The spatial plan of Legler emphasizes privacy and seclusion. The development is dominated by bi-level and ranch homes on large lots. During a visit in 1981, I saw swimming pools in many backyards. Stores and other amenities, where neighbors might meet each other, were absent from the area. People were dependent upon cars for meeting almost all needs. There was little visual evidence of community; a firehouse was the only visible tie to Jackson Township.

Legler was only sparsely populated prior to the late 1960s. Single-family houses lining the highway attest to an initial period of growth from 1968 through 1971. A similar growth spurt from 1973 through 1975 began to fill in a road running parallel to the highway. A final and major surge of development that pushed back from the major roads began in 1976 and peaked in 1978, virtually halting by 1979.

The new residents came primarily from dense urban areas of New Jersey or from New York City, where they tended to have had a surprisingly stable pattern of residence. Half the sample had previously owned a home; but in contrast to these "starter houses," all considered their Legler house to be a "permanent home." The move to Legler often reflected more than just having outgrown the prior house. It also involved an escape from a hectic urban lifestyle in crowded neighborhoods. The dominant factor cited in the selection of Legler, mentioned in some form by nearly everyone in the sample, was its rural ambience. Legler was viewed as being undeveloped and isolated—it had woods, fresh air, well water, open space, sparse population, little traffic, and quiet. Work-related considerations (e.g., commuting distance, job transfers, or room to park one's truck) were also cited by nearly half of those interviewed.

Typically, the search for a new home coincided with some milestone, such as marriage, starting a family, outgrowing the prior house, or achieving the savings necessary to afford moving beyond a starter house.

The most common strategy for selecting the new home involved a general search that targeted Legler only after the family saw an advertisement or visited a builder's model. For some half of the sample, the move to Legler was based primarily on financial considerations. Legler was a comparatively inexpensive place to acquire an acre of land and to build a custom home. Respondents stressed that they had wanted "new" homes. The $40,000 to $50,000 price range was considered a bargain during the peak years of development in the area.

Residents used words such as "Shangri-la," "countrified," "beautiful and peaceful," and "clean and refreshing" to describe their recollection of Legler at the time they moved there. The country was seen as a place where children, free of the perils of crime, traffic, and "Cancer Alley,"[4] could have wonderful adventures that many of their parents had never experienced. In nearly all cases, it was seen as a generally healthy place compared with the new residents' prior homes.[5]

Privacy and Community in Legler. The physical environment changed as the area was developed but there were also interesting changes in the social environment. For residents who predated the period of intensive housing development, Legler was sparsely populated and isolated. Residents depended upon each other for social stimulation and developed a tight self-help network as well. This is expressed by an old-timer, who recalled that

> *Everybody was friends. They were the best, most beautiful people; there were no enemies. We had parties at Halloween, Thanksgiving, on every holiday. On Christmas we'd ride through the area on horse and buggy. If anyone was in trouble, everybody would help.*

Although a feeling of closeness permeated the area, a strong sense of reserve and respect for each family's privacy was also evident.

> *It was a quiet, no pressure place. We weren't coffeeklatch people. We were here for our neighbors if needed, but we didn't want people visiting all the time.*

During the first major influx of development, these old-timers extended help to their new neighbors, inviting them over for a drink or assisting them in the move. They found that the newcomers shared the old-timers' rural expectations for the area even if they lacked the communal spirit that had previously characterized Legler. Proximity was the newcomers' sole basis for association. They established a "hello-basis" for neighboring and helped each other readily. However, they did not seek to establish close ties with their fellow residents.

We felt cut off from people and that's how we liked it.

In selecting Legler, the new residents had deliberately avoided "housing developments." Given the way they valued solitude and a rural way of life, it is not surprising that they wanted to shut the door to Legler. Accordingly, they resented the predominantly young families with children who came several years later and who constituted the largest and final phase of development in Legler (from 1976 to 1978). During the interviews, earlier residents derided these latecomers for their suburban "coffee klatch" relationships. In fact, if reserve and solitude were the main components of privacy for the earlier residents, intimacy and neighboring characterized this later group—their expectations favored a suburban rather than rural way of life. Given the difference in residential values between these groups of residents, it is not surprising that they had little contact prior to the pollution incident.

For the latecomers, social relationships were initially affected by the pace of development at the time a family moved in. While it might be lonely for the first family in a new subdivision, strong local ties quickly developed once neighbors arrived.

We had nice neighbors in our age group with children similar in ages to ours. We had met the people when we checked on the house before we moved in. The whole line of six houses is very closeknit. We picnic, share a garden and let the kids play together. Our neighbors are either here or we're there. We share a pool and party together almost nightly. Given that we all have seasonal jobs, we even help each other out when we're having a slow period.

Elderly residents, adrift in the sea of young couples with children, came to play a role as honorary grandparents. Not every section of Legler was so cohesive. In one area, settled largely by people from outside the region, problems with the builder led to tension among the families that continued until the pollution incident began. Other influences also affected the social environment. Residents who had moved from nearby areas retained their friends outside of Legler and thus felt less pressure to relate to their neighbors. The overall social dynamics of the community reveal that the closest friendships were based upon proximity—until the water crisis, when people from across Legler began to relate as a community.

In fact, Legler's status as a community was very nebulous prior to the drawing of pollution boundaries. Residents arriving during the period of development tended to have a very vague conception of "Legler" altogether. Many knew little about the area before moving there. Jackson was seen as a sprawling town lacking either a center or orienting landmarks. For earlier residents, the Legler section was viewed as the "boonies," the rural, undeveloped part of town. But given the reports of real estate agents, later residents, with their suburban orientations, expected the section to continue to develop into a more urban center with shopping facilities. As a further indication of the limited sense of community, only one respondent reported having joined a community organization (the fire company) that gave him a basis of association with fellow residents.

Disrupting the Dream

Reflecting upon the time when they moved to Legler, respondents recalled seeing no major drawbacks to their new neighborhood. The only concerns noted involved commuting distance, distance from shopping, and problems with the builders. In virtually all other ways, Legler was the "American dream" achieved. But then, things began to change. The origins of these changes predated the arrival of most of the residents.

The Glidden Mine Site. In 1961, the Glidden Corporation began mining on a 135-acre site in Legler for a mineral used in manufacturing paint. Although many local families were unaware of the operation, families near the mine were heavily affected. A key local road was severed by the mine. One family's home, surrounded on three sides by the sandy soil left over from the operation, came to be called "the island." The mine site itself acquired the name "2001" because of its moonscapelike terrain. Area teenagers later frequented the site at night, swimming in the water-filled craters. Despite the aesthetic loss to their neighborhood from the mine, residents appreciated the jobs it brought to the area. Additionally, they were led to believe that the site would eventually be reclaimed as a park and the truncated road reconstructed. Picturing a rosy future, they were tolerant of the operation.

A Landfill for a Neighbor. However, in 1971, when Glidden faced the prospect of closing the site and undertaking an expensive reclamation job, Jackson was seeking a new site for its municipal landfill. In a true marriage of convenience, the two realized that they both could benefit if Glidden deeded the site (for one dollar) to Jackson. When Legler residents discovered that their neighborhood was being considered for a dumpsite, they were alarmed. There were several reasons for this concern. First, residents had long awaited reclamation of the site. Second, they did not want garbage from the entire town dumped in their area, especially when they felt that services were concentrated on the other side of Jackson and the Legler area did not get its share of benefits. Third, residents realized that the landfill might threaten the peaceful ambience of their neighborhood.

We didn't want it because Lakehurst was a country road, it was quiet. We didn't want to go from an isolated area to Grand Central Station.

But the greatest concern was about pollution. The sandy site was full of sixty-foot-deep holes filled with water. The wells of surrounding residents tended to be shallow—often less than fifty feet. Moreover, the dump across town was being closed because it had caused water pollution, it smelled, and it was a breeding ground for rats. Legler residents foresaw having the same problems in their area.

A group of Legler homeowners quickly organized to oppose the landfill. A petition with 365 signatures against the Legler site was presented to the township. Many residents attended meetings to voice their protest. In the view of my respondents, however, the decision had already been made in secret. It was too late.

The township went to great lengths to reassure residents that the new landfill would be run properly.

People believed the promises, for example, that there would be gate passes and dump tickets and other safeguards, that someone would carefully watch the dumping and that there would be security.

Residents were also told that after ten years the landfill would become a golf course, enhancing their properties. Despite the failure of Glidden to keep a similar promise, some residents accepted this promise as well.

After their unsuccessful protest, many of the residents felt bitterness toward the township. They felt that the landfill siting decision had been manipulated. Furthermore, the township's promises quickly proved to be short-lived. Yet, not seeing any other course of action to pursue, residents ceased their organizational work against the landfill and accepted it as a reality. As one resident explained, "We didn't like it, but we had no choice. You can't fight city hall."

This incident was important in several ways to later developments in Legler. People who had never before mobilized for a cause learned to be community activists. They also learned that local officials were not to be trusted. While no general community organization emerged for some time after this, a concerted effort by the Legler activists to unseat the dominant local party in the 1972 township election was successful.

The municipal landfill operated in Legler from late 1971 until late 1980. As many as 50,000 gallons per day of human waste were deposited in the landfill through 1978. The operation caused ongoing controversy over controls for permitted dumping.

Although earlier residents had been greatly concerned about the landfill, many of those arriving after the landfill began operation were unaware of its existence, even when the homes they purchased bordered the facility. Others, apprised of this situation, were surprisingly unconcerned. That these residents could be so completely uninformed is attributable to several factors. First, people did not explore their future neighborhood for possible hazards; instead they assumed it was safe. Second, when they learned about the landfill, most people were convinced that the facility was not only benign, but some even viewed it as a convenience. Residents pictured the landfill as a sand pit or a local dump for residential garbage. Many of the residents had moved from areas where it was common to build upon landfilled ground. They appreciated neither the scale of the landfill operation nor the range of wastes accepted. Finally, misinformation was apparently deliberately given to residents by real estate agents, lawyers, and township officials, as indicated by this comment:

> We never walked to the back of our property because of the weeds. I did ask the real estate agent what was behind the property. He said that there was a closed landfill which had just been for local garbage and that the property was going to become a park. We didn't give it a second thought. We didn't even know what a landfill was. We figured it was a pit that you take sand out of. After the first week in the house, I took a look at the back of the property. They were dumping right at the end of our lot—tons of garbage.

Some new residents were mollified by town officials who told them that there was a greater chance of well pollution from their septic tanks than from the landfill. Others, living as close as one-quarter mile and as far away as three miles, believed that the landfill was too far away from their homes to affect them. Thus, even when they discovered that they had been deceived about it, most newer residents rationalized the presence of the landfill. In contrast to the begrudging acceptance by residents who preceded the landfill, the facility just was not a concern for the latecomers. Thus, for most residents—new or old—the Legler facility settled into the background of the neighborhood.

The Direct Impacts from the Landfill

Although the landfill came generally to be an accepted fact of the Legler section, there were constant reminders of the operation for residents exposed to its direct effects (see Edelstein, 1980). For some strongly impacted families, the facility contributed substantial life stress long before it was labeled as a polluter. Figure 2.1 shows the approximate distribution of direct impacts within the Legler area, illustrating the situation faced by Legler residents during the period when the issue of water contamination was still in incubation. I will briefly review some of the key direct impacts.

Odor. Odor was one of the worst impacts for those living near the landfill, as these comments suggest:

> *Depending upon which way the wind blew, we would get the odor. We had built a patio prior to the landfill's opening so that we could use our backyard. By the following summer, we couldn't stay there. We put in air conditioning in the house because we couldn't open the windows in the summer.*

> *We couldn't hang out our clothes on the line; the smell would get to them.*

> *The smell came right through the house. Sometimes you'd wake up in the middle of the night and think you were in a sewer. If you'd start a barbecue, the smell would come and spoil it. Our children couldn't have company or sleep out at night.*

Noise. Facility noise comes from both traffic and equipment. In the case of landfills, heavy equipment is used to move the dirt needed to cover the garbage. Such equipment operates at high-decibel levels that destroy peace and quiet. Even if a nearby resident grows accustomed to the rhythm of this equipment, intermittent irruptive noises are also produced, such as the "beep-beep-beep" emitted when equipment is shifted into reverse. Such noises are unpredictable for listeners, demanding their attention. These noises are particularly stressful (Klausner, 1971; Reim et al., 1971). Truck traffic into landfills is associated with noises that include the constant roar of engines, the sounds of shifting and braking, beeping of horns, and the banging sound of the metal bodies on the rough roads.

Traffic. Typical concerns with traffic in residential areas include safety, intrusion on local roads by cut-through traffic, noise, congestion, litter, and air pollution (Edelstein et al., 1975). Traffic impacts extend along all routes used to reach a waste site. Landfills also bring congestion, particularly in the morning and at times when the public has access. Traffic safety concerns are often focused upon children, as this Legler resident indicated:

> *We worried about the kids on the road. We had to curtail their activity. We wouldn't allow them to ride their bikes until after the landfill closed for the night.*

Additionally, neighbors of landfills report that idling trucks fill the air with fumes.

Litter and Dust. Blowing refuse and dust are frequent problems at landfills. Litter results when trucks are inadequately sealed or when

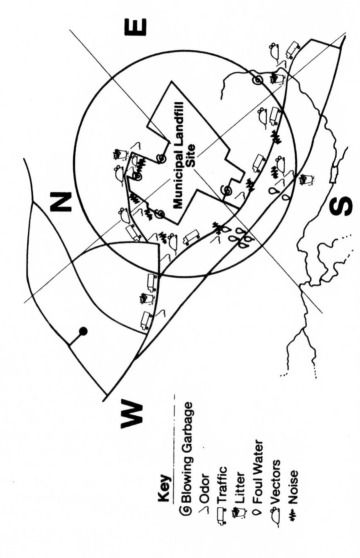

Figure 2.1   Direct Impacts from Municipal Landfill, Legler

deliberate dumping occurs along roadways, usually when a driver finds the facility unexpectedly closed. In Legler, properties adjoining the landfill were covered by blowing refuse, as this resident recalled:

> *First people would dump in front of my house and then litter would blow onto the back of my property. I was surrounded by garbage. The garbage in the back was even floating in my stream. The township agreed to clean it up but they didn't. Later the health department ordered me to do it. I took three truckloads out.*

At other landfill sites, "Saharalike" conditions have also been described, with dust coating the area and entering nearby houses, in one case causing respiratory problems.

Other Impacts. Additional direct impacts include those from loss of aesthetics and from fire, explosion, and corrosion. In Legler, as elsewhere, vectors were associated with landfilling.

> *We had mosquitos—we never had them before the landfill nor since it closed. Also we had rats in the yard. The township poisoned them. We feared small pox from the sludge lagoon transmitted to the children by animals.*

Finally, residents for whom reminders of the landfill brought bitter memories found additional disturbance from the meaning that they associated with the constant intrusions from the facility. In addition to odors, noise, and litter, they confronted the realization that unwanted changes had been forced upon a community that otherwise they had seen as ideal.

Portents of Disaster

Although these direct impacts were stressful, Legler residents generally learned to live with the inconveniences. They sensed that there was little that could be done to stop such impacts once the facility was permitted to operate. For them to become active in opposing the landfill at this point demanded some major event, such as a particularly frightening incident or notification of contamination. Ironically, the period of incubation is often accompanied by numerous clues that disaster is brewing, yet the concern generated by such clues is below the threshold for action.

In Legler, prior to the recognition of the contamination, seemingly random water quality problems had appeared. As with health problems that had also affected some area families, these tended to be treated as private problems, with neighbors possibly unaware of each other's concerns.

> *One month after we moved in, we thought that the water was funny. We spent $1,300 on a water purifier. Soon it was bad again. We talked to another person with this problem. They didn't feel good either. We had itchy, dry skin.*

> *Every time we'd drink coffee, we'd immediately have to go. The water had the worst smell. You had to open the windows. You couldn't run the water in the kitchen.*

In some cases, one member of a family perceived a change in water quality that escaped the notice of other family members.

There were also indications that landfill security was inadequate. Thus, Legler residents reported,

> *We took pictures of them trying to cover 300,000 tons a day of garbage with sand. It didn't work.*

> *I saw them dump 43 barrels of chemicals. I tried to move one of the 55-gallon drums, and I couldn't. Then I noticed that something had eaten a hole right through my pants.*

> *I saw green and purple colored sludge in pits thirty feet deep. I didn't recognize this as a problem at this time.*

> *I've dropped garbage off and seen pools of water with garbage down to the water table. I didn't realize that our well was so shallow. It's easy to connect the two—I saw it with my own eyes.*

> *I had a friend who was offered money to keep quiet about a truck that was not licensed to dump there.*

That residents could view these clues and yet not recognize the pattern suggests a frightening aspect of lifestyle and lifescape accommodation— the rejection of anomalies that undermine basic assumptions about life. While Legler residents found the health and water problems and indications of poor landfill security disturbing, they apparently did not comprehend the significance of these events as signposts of toxic contamination.

The Legler contamination came to light in November 1978, only a few months after the Love Canal debacle hit the press. And yet nearly half of the Legler residents interviewed either claimed never to have heard of toxic pollution previously or else they ignored it. Others recall hearing of incidents and thinking that it could never happen to them. The awareness of the landfill, whether among old-timers who had accepted it or among the newcomers who barely knew of its existence, was not associated with the dangers of toxic pollution. As two Legler residents reported:

We had heard of Love Canal, but it wasn't really relevant to here. We were two miles from the landfill; nothing could happen to us.

We heard of Love Canal on the news. I felt sorry for those people, but I didn't lose any sleep over it. We didn't think it could happen to us.

Caught up in their families, work, home, and the other concerns of the American lifestyle, Legler residents did not expect what was coming despite clues that would serve as the basis for explanations after the fact. What was about to happen to them was, at that time, virtually inconceivable. Thus, after an initial period of incubation, during which they accommodated in various ways to sources of stress associated with the landfill, Legler residents entered the next phase of toxic exposure, the period of discovery and announcement.

Discovery and Announcement

Unbeknownst to most residents, the New Jersey Department of Environmental Protection had responded in the summer of 1978 to a request from the Jackson Township Board of Health to test Legler's groundwater, part of the vast Cohansey Aquifer. Although no pollution was found, residents' complaints about water quality continued. In the fall, the DEP did further testing for contamination by organic chemicals. In October, the first indication of significant pollution was found (Division of Epidemiology and Disease Control, 1980).[6]

In response to the October discovery of organic chemicals, on November 8, 1978, the Jackson Board of Health issued a notification to residents south of the landfill that their water was polluted. Referring to the board meeting held to review the test results, the notice read in part: "It was recommended that because there was enough evidence for concern, area residents are advised *not* to use water for human consumption, but limit use for sanitary purposes only." Later versions of the message additionally recommended that residents ought "not to soak for extended periods of time" when bathing. With this notification, there commenced for Legler families a period of disruption that has had continuing ramifications.

Notification was originally given for only a portion of Legler. Others outside the original boundary were reassured that their water was drinkable. But over several weeks, official identification of the area affected by pollution was changed, so that by early 1979 all of Legler's approximately 150 families had received the warning. It was ironic that the identified boundaries of contamination so closely mirrored Legler's own boundaries. Not having previously heard the label "Legler," most residents only became aware of their neighborhood's identity with the advent of the disaster.

Reactions to Notification

In many cases, residents were notified in person by a representative of Jackson Township. Sometimes the messenger's demeanor intensified the warning in the letter. Thus, one respondent recalled, "I knew from the look on his face that it must be serious." Other people, however, returned home merely to find a form letter left for them.

Initial Groping. How did people react to notification? In moving their young families to the suburbs, most Legler residents had just invested their hopes and their savings in new homes. A number had just been issued a certificate of occupancy (CO) certifying that their new home could be legally occupied. With the announcement of contamination, nearly all were caught by surprise.

> *Four days after our Certificate of Occupancy was issued, we were eating dinner with our neighbors. A man came to the door and announced that we might get cancer if we drank the water. Our neighbor refused to finish her spaghetti because it had been cooked in our water. She was quite sick. We couldn't believe it—four days after our CO! What were we supposed to do now?*

Initially, people were confused by the notification. Women called their husbands at work to relay the news. People frequently called neighbors to see if they too had been notified and to find out what they were doing. Many residents also called the township for clarification. This latter step did not necessarily prove helpful, as illustrated by one resident who was told "It's too technical for you to understand."

Reactions to the notification varied. There were some indications of initial panic. One resident spoke of "going berserk." Others viewed notification as an attack and called the police. Some residents remained calm in the face of notification. Others believed, at least initially, that this was a "problem" that could simply be fixed.

> *My husband was calm. "It's no big problem, they'll fix it. It's like fixing a water main that breaks." Then we went to a township committee meeting and we learned fast. We had expected this to be fixed in a couple of weeks.*

Alternatively, other residents minimized the risk associated with the chemical exposure.

Nothing bothers me. I hate to sound blasé, but I see in the paper something that causes cancer almost every single day.

One resident, her family already reeling from a series of disasters, commented:

This was just another blow in a long line of blows—we were dumbfounded.

Trying to Define the Uncertainty. As they began to assess their predicament, many residents turned to their elected officials for guidance, only to be frustrated with the township's inability to clarify their situation.

When I asked at a meeting how long before we would get clean water, the health department official didn't know. There was no light at the end of the tunnel.

Residents began to realize that their elected officals didn't know how to remedy the situation. The officials responded to the uncertainty with extreme caution, causing concerned citizens to feel abandoned in a state of extreme uncertainty.

This was like someone shooting you and then not having the decency to finish you off.

To Drink or Not to Drink. As they sought to understand their new situation, Legler residents had to decide whether to discontinue the use of their well water. Most of them ceased using it immediately upon notification. Some made transitional steps toward stopping use. The transition was hard, with people falling back on long standing habits.

I would go to the faucet and fill up a pot and then remember and have to spill it out.

Most residents continued to shower in the water, and some used it to bathe. Generally, parents were more careful in limiting their children's exposure to the water than they were with themselves. Most discontinued use of the water in cooking, although some apparently believed that the water was safe if boiled. Some residents continued using the water for an interim period until they became convinced that it was polluted. A few others continued to use the well water until forced to join the local water district several years later.

Rising Frustrations. As they were left in uncertainty, residents increasingly directed their anger at the township. From the beginning, residents were enraged to have been put in this situation. Earlier residents were angry that the township had forced the landfill on them and then not operated it carefully. Recent newcomers had a different historical perspective. They wondered how the township had approved certificates of occupancy for their houses, making the new owner rather than the developer responsible for addressing the water problem. Now these residents feared the loss of their COs due to the contamination.

> *My mother had read about the pollution in Legler. The builder said there was no problem. Two weeks after we moved in, I called the Board of Health to make sure our water was okay. They said there was no problem. Later that same day, the notice was delivered that we couldn't drink the water. I was angry at the builder and the township over the CO. The builder had lied that he lost the water tests to give him time to finagle with the town to let us move in.*

The degree of anger resulting from this situation was indicated by another respondent.

> *I went to three meetings to ask the mayor how I got my certificate of occupancy after they knew about the pollution. Each time I got so mad that I couldn't speak, I just walked right out.*

Whether people actively believed that the water was polluted or not, they were quick to appreciate the implications of losing their COs, being unable to sell their homes, and having to finance a new water source. Young families had invested every penny in their homes; older residents had limited financial resources.

> *At first we thought we'd abandon the house, but everything we owned was sunk into it. We were in debt. We were trapped. There was no place to go. No one wants to buy Love Canal.*

> *We were all set to sell our house. We were supposed to sign the next day. We lost our CO; we lost our sale. Our concern wasn't health—we didn't know about this yet. Our investment was destroyed.*

A major by-product from the lack of an easy escape was that residents were forced to face the problem head-on.

Developing a Social Consensus. Beyond the lifestyle impacts of forgoing tap water, however, there were actions necessary to clarify an uncertain situation and to bring about its resolution. When it became apparent that Legler residents could not rely on local government to provide a timely and adequate remedy, people were forced to adopt increasingly sophisticated means of involvement. As Legler residents met at township meetings, they began to communicate with each other. From their shared frustration, revealed during the confrontations with the township council, came the potential for local cooperation. Under the leadership of community residents who had initially fought the landfill, an organization called the Concerned Citizens Committee of Legler was created. The meetings of this organization played the vital role of creating a tangible bond, a way for this community in distress to come together to confront the issues.

Nearly all residents interviewed saw the notification of pollution as pulling people together. There had been an initial lag. Old-timers had organized among themselves quickly, while newcomers, who lacked local acquaintances, were left to fend for themselves. But as the organization began to develop strategies requiring broad participation, it recruited the new residents. People met with others whom they might never have known otherwise. As one newer resident observed,

> *This pulled a lot of people together at meetings. We met a lot of people. We had concern for each other. Sympathy. We felt bad for people who lived here a long time when we saw what they'd gone through.*

As people became more involved in the process of discovery, they encountered a growing social consensus regarding the severity of their situation. Thus, even if they were initially reluctant to believe the notification, they were subjected to a milieu in which continued denial became increasingly difficult. Furthermore, over time, more wells evidenced pollution and more information about the landfill came to light.

Substantiating the Threat. A number of developments contributed to the growing consensus within Legler about the nature of the problem. The informational meetings held by the Concerned Citizens Committee provided the first firm information. At one influential early meeting, a geologist from the Department of Environmental Protection (DEP) spoke frankly about the dangers associated with the chemicals that had been found. This straightforward information challenged the skepticism felt by many of the residents about how serious their situation really was. As one resident recalled,

> *I didn't view this as a serious problem until the DEP meeting with Concerned Citizens. Then I knew we were in big trouble. When I heard about the chemicals in the water, I felt sick.*

In addition, some residents received results from a state groundwater study of their wells, ending any misconceptions about the water's purity.

When we found things like benzene in the water, I was shocked, worried and frustrated. I thought the test would show that we were clear.

Residents having technical expertise were particularly quick to grasp the severity of the situation.

Also contributing to the emerging consensus was media coverage that took the situation very seriously, reinforcing the fears of some residents and adding additional information about the implications of toxic exposure. As one couple recalled,

We heard about the cancer-causing chemicals on the news. We worried about the kids. They were scared too. They asked us questions about cancer.

Finally, there were continuing reminders of health problems. After notification, some people exhibited new symptoms, such as rashes and skin irritations associated with showering. As people began to communicate with each other, they compared symptoms, reinforcing the view that their problems derived from a common source, the water. People then began to rethink previous health problems and to link them to the consumption of the water.

We mothers began to compare notes and decided that the ear infection and other problems with our children were no coincidence. I saw our family as being in jeopardy.

You hear so much about the water. I heard that benzene can be absorbed through the skin, yet the town told us we could still shower. I heard about cancer clusters and the kidney problems. I was scared.

On the day after notification, Jackson Township began to supply water to residents in Legler. Later, planning began for a central water system to replace residential wells. For some residents, it took this level of response to finally convince them that the crisis was real.

Disruption of Lifestyle—The Delivery of Water

Shortly after the first notification, water tankers were placed at both ends of one of the main roads through Legler. After Liquid Plumber was inexplicably found dumped in the outdoor tanks, the township began to use large metal civil defense containers to deliver water to each of the affected households. When a family required more water, the resident signaled by placing a flag in front of the house. Protests over the clumsy and unsanitary metal containers led to the adoption of large plastic containers, each outfitted with a spigot. For some two years, until their homes were connected to a central water main, most families received all their drinking water through deliveries. At the time of the interviews, in the summer of 1981, several families were still drinking delivered water.

As many Legler residents pointed out, Americans assume that they will have a ready supply of usable water at their fingertips.

> *Water is a life source. We lived in the city. You just had to turn the tap on. We didn't think of these things. We never thought about it.*

The way that we organize our homes, our schedules, and our activities depends upon a ready supply of water. When a disruption of this supply occurs, the ramifications are felt as a disruption of lifestyle itself.

Impacts of Water Delivery

Various impacts were associated with the inability of residents to use their tap water. While most respondents spoke of the period of water delivery as one of inconvenience rather than difficulty, it is evident that this was a stressful time for residents. It was easy for people to quarrel. Daily life became a chore. A general sense of well-being was hard to maintain.

Water Containers. The logistics of getting water into the home disrupted lifestyle during this time. The containers were heavy and difficult to bring inside, contributing to secondary health impacts and family friction.

> *My wife would carry the 17-gallon jugs. She did nerve damage and got disc problems in her back. Now she has to take it easy.*

> *I was not able to carry the water because of my health problems. This hurt my ego. And my wife was mad because she had to do it. We'd argue about this.*

Water Quality Issues. There was also concern over the quality of the delivered water. Some residents refused to drink the delivered water because it tasted so much worse to them than did the polluted well water. Additionally, the delivered water might be muddy or over-chlorinated. It might contain cigarette butts, aluminum, sand, or, after the township repainted the containers, paint chips. One woman pulled a big jar of water out of her cupboard to show me. At the time of delivery, the water had been slimy. More than a year later, the solids had separated out to reveal a quarter inch of black particles at the bottom. Finally, people who argued with the delivery workers claimed that their water was at times tainted with urine.

Water Availability. Families also had to manage water use carefully to ensure that an adequate supply was on hand. There were also times when trucks broke down or delivery was not made for some other reason, causing the delivery to be viewed as unreliable. Also, to arrange delivery, people had to remember to put their flags outside the night before.

> *Having trucked water adds responsibility. It's something you have to keep in mind all the time. If we run low, there's no way to get any on the weekends. We have to ration on Saturday and Sunday.*

People always had to judge how much water they had and how much they needed. Too much water might stagnate. Too little water had obvious shortcomings. And if no delivery was made, residents had to call town officials who might blame them for failing to put out the flag, even if that had been done. Winter brought another problem. The cans of water froze if left outside, creating huge ice cubes capable of splitting the container and, upon thawing indoors, flooding the floor.

The Pioneer Life. Even when people mastered the logistics of using delivered water, they still had to contend with the difficulties of not having tap water in the home for housework and for personal and recreational use. A number of residents characterized their ordeal as "the pioneer life." Although some people had access to alternate washing facilities (e.g., at the home of friends or at work), most residents learned to live with minimal bathing. Most utilized the tap water for showers. One family whose shower water tended to burn their eyes learned never to face the shower stream. Later, after their home was hooked into the central water system, they still maintained their aversion to facing the water. Some adults who had severe skin reactions from the water bathed with the delivered water. While several of them heated the water in pots on the stove, one woman reported washing in a barrel of cold water each morning. Invitations were extended for residents to shower at a local school. However, few availed themselves of this opportunity. They felt that winter was an unhealthy time to shower prior to being outside. Furthermore, people were reluctant to use an open shower room; it wasn't private enough.

People had to remember to carry a cup of water from the storage can to brush their teeth. This was one task about which people reported "cheating" and using the tap water.

Homes often were not kept as clean as they previously were. People varied between using the tap water, the delivered water, or no water in cleaning. Housework took much longer.

> *I'd be home alone. I'd try to bathe the baby, cook and clean with the delivered water. I'd live out of a jug. It was a real strain. I have a bad back; it was impossible for me to carry the water to the tub. It would take longer to do anything using the water. It was inconvenient.*

Many young parents bathed their children in the delivered water. This entailed carrying water from the garage, heating it on the stove, and then carrying it to the tub. Most people continued washing clothing and dishes in the well water. A few even used it in cooking. With delivered water, cooking became a major chore. Vegetables had to be soaked to clean them.

> *If you were making spaghetti, you couldn't rinse it. You'd have to pour water over it. It was aggravating, especially because this wasn't your fault. Who needs this?*

Families with animals had to hand-carry water for them as well.

Recreational use of the property was also affected. It was more difficult for residents to use their backyard pools during this period. Some residents who had planned to build pools delayed until after the central water system was installed. Gardeners were generally forced to depend upon rainfall, although some had been told that aerating the water made it safe for vegetables. Some gardeners gave up the hobby.

It also became harder to have guests during this time.

> *What happens if you have thirty people coming for dinner and you have no water?*

> *We put big signs on the faucets—"Don't drink the water." It was embarrassing.*

> *People would bring water with them when they came to visit instead of bringing cake.*

As these examples illustrate, the period without tap water was one of constant personal frustration. As they went about the small steps of daily life, residents were further reminded about the fundamental crisis in which they had been trapped.

A Dominating Concern. Beyond these complications and disruptions of routine, a further change in lifestyle occurred. For many victims, the exposure incident became a central focus of their lives, dominating free-time activity; conversations with neighbors, friends, and relatives; and family life itself. The problems faced by the victims required making many decisions, such as whether to stay in the home or what degree of exposure to the tap water was allowable (for example, should the children play underneath a sprinkler?). At the same time, residents' understanding of the situation was limited because of a paucity of information, the complexity of issues about which they had little prior experience, and distortion by the media or by agencies anxious to minimize the problem.

These impacts were compounded by secondary effects. For example, as a resident began to devote greater attention to the pollution incident, the person's family life, friendships, work, and recreation were adversely affected. Friends and relatives grew tired of the victim's obsession with the incident. Spouses became resentful at the absence of a mate active in responding to the contamination situation. These issues will be more fully explored in later chapters.

The Hookup of City Water

Beginning in July 1980, Legler residents were able to connect their houses to the central water line installed by the township. This water district used the deep well beneath the Glidden mine site as its source. Given the lifestyle disruptions that characterized the period of water delivery, it is not surprising that having their homes hooked up to the central water system was a major event for most Legler families. Water had been transformed from a resource assumed and taken for granted into a prized and valued possession.

> *The water was like gold. We let it run; we drank it. We all cheered; we were elated.*

> *We had a party in the neighborhood when the water came in. We ran five hoses [one from each neighbor's house] to fill our pool. We filled it in four hours. The whole neighborhood jumped into the water that night. We didn't drink beer for once, we drank water.*

At the same time, it was not so easy for some residents to forget the legacy associated with their water supply. There was bitterness in this remembrance.

> *I broke the well pipe when we put in the city water. I just felt like locking the front door and leaving, like in The Amityville Horror.*

Not all families initially availed themselves of the opportunity to join the water district. By the summer of 1981, twenty-one of twenty-five families interviewed had hooked into the system. The others were unable or unwilling to pay for the connection but were later forced by the township to join.

One might assume that having a new water source would end the disaster. But only one respondent indicated that having the central water ended the person's concern with toxic exposure. It is an important lesson of toxic disaster that its scars are not easily mitigated; replacing what was lost in Legler involved more than obtaining a different source of water. Some losses seem beyond mitigation.

For the Legler residents, several specific concerns remained. Sensitized to the pollution of the aquifer beneath them, residents feared that the well at the Glidden mine site, although deep, would eventually be contaminated by chemicals migrating from the adjacent landfill through underground water. One resident revealed the general anxiety about this issue when he recalled trying to second-guess the township.

> *I just hope and pray that this water is safe. You never know. A worker came and told me recently that he was taking a test for chemicals. I got upset and asked why. When I called the township, they said it was just a bacteria test. But why should they take this from the tap?*

The same resident also remembered one of the township's arguments against the digging of new private wells.

> *When the town talked to us about deep wells, they explained that you need double casing to prevent chemicals from getting into the water. Well, does Glidden have double casing?*

Furthermore, the water system was run by the same township that operated the landfill that had polluted their wells. In noting that this concerned them, several people referred to the history of distrust toward the township, as this comment reflects:

> *We suspect the new well. We don't trust the town. They're the same people who pulled this stunt to begin with.*

An additional concern was the quality of the water, perceived to be excessively chlorinated. One resident noted that his family was still buying bottled water. Another said,

> *It tastes yucky and disgusting. It's like drinking Clorox out of the bottle. We boil it and refrigerate it.*

Others indicated that their concern for chlorine was more than a matter of taste. They were concerned that chlorine is itself toxic. Further adding to water quality concerns was the use of asbestos pipes, viewed as a potential health threat, in connecting the water system.

> *How will we know when the pipe eats away?*
>
> *We get killed either way—if not from the water, then from the pipe.*

Residents' fears were later substantiated when tests revealed detectable amounts of asbestos in their new water source. While no follow-up interviews have been done since this discovery, it is safe to assume that it fed the fears and distrust noted here.

Another major issue was that the water system was now seen as being out of the control of residents. The water was viewed as unreliable. One resident recalled his anger at being stuck lathered in the shower one day during a service disruption. Fueling these concerns was the townships's approach to citizens during the hookup phase. Frustration mounted because "they didn't listen to people." Residents reported the heavy-handed manner by which the township forced recalcitrant residents to hook into the system by threatening to condemn their houses if they did not. Old-timers on fixed incomes were most angry over this practice. They minimized the threat of drinking the well water, and therefore saw no great advantage to hooking into the new water system. As one noted,

> *The town says that I have to connect to the water! Why? What law did I break? We own a well. They have not proven to us that the water is contaminated. There is no solid evidence of a problem.*

Additionally, the hookup represented a major financial burden for some residents. Once the township provided the central water main, it was up to each household to pay a hookup fee and to connect into it. If the house was a long distance from the main, the costs of trenching and burying the pipe were high. A few families could not initially afford the connection. One family sold its horse to pay for it. Others borrowed money. A number of residents dug their own trenches for the hookup. In some sections, there was a cooperative effort among neighbors. These residents told of various roadblocks to their work, including trenches that collapsed and inspectors who demanded deeper ditches.

Once hooked up, residents then had to pay for the water they used. This additional financial burden irritated residents, who perceived themselves as double-taxed because they had already paid for wells and now had to pay for the water system. Compounding the anger was their view that rates in the Legler water district were high compared with those in other parts of the township. The costs proved a barrier for some.

> *We paid the hookup fee, but we haven't plumbed in the water. We're on social security—fixed income. We can't afford to pay thirty dollars a month for water. This would be an awful burden on our income.*

Finally, beyond the possible contamination of the new water source, residents also feared other avenues for toxic exposure from the closed landfill, such as through soil, leaky basements, and surface water. They also feared that toxic residues might persist in their plumbing.

> *I figure it's in the ground. I wonder if they dumped it near our wild blueberries.*

> *I still have a well; I'm still in the affected area; the chemicals are in the air and ground—they could still ooze out! I think about it in the summer when I'm gardening. Will the ground contaminate the plants?*

In addition to problems that persisted in the environment, residents had concerns about damage already done to themselves and their children. Thus, whether the new water was good or not failed to obviate the concerns raised by the fact that the old water was bad. As one resident remarked, "Cancer does not show up right away." Another referred to this persistent health theme:

> *We still worry about our children. It will always be on our minds for a while yet, until they grow up.*

These ongoing concerns demonstrate that a new water supply represented only one step toward reestablishing some sense of normalcy in Legler. Although lifestyle disruptions generally ceased as a result of the hookup, return to normal lifestyle did not end the disaster. Many of the residents have undergone profound changes, changes that are deeper and more permanent than those that accompany disrupted lifestyles. The persistent changes involve the lifescape, or the way that residents view their world. It is to these lifescape impacts of toxic exposure that I turn next.

Notes

1. For a detailed review of the study methodology, the reader is referred to Edelstein, 1982, available from the author. All quotes pertaining to the Legler case study are drawn from this source.

2. In nearly all cases, I spoke with both adults in the home. Children were often present as well.

3. In 1981 Legler was demographically similar to Jackson Township, whose population of 25,000 was almost entirely white, about equally divided between men and women, and heavily skewed in age toward younger residents. The modal income range in Jackson in 1970 was between $10,000 and $14,999. At that time there was an average of 3.79 people per household. In 1979 all units built were single-family dwellings. Jackson's population statistics are subject to substantial increases (estimated at one-third) in the summer to account for tourism and seasonal residents (Ocean County, 1981).

4. Dr. Adeline Levine was extremely helpful in sharing her early Love Canal results during the development of my study. She had already found age and the presence of children to be influential variables in her Love Canal research.

5. "Cancer Alley" was a term that came into general use in the New York metropolitan area in the mid-1970s. The reference was to the industrial center and northern section of New Jersey, where elevated levels of cancer had been identified by the National Cancer Institute (see Greenberg, 1986, and Greenberg et al., 1980).

6. Testing continued for another eight months. As many as 34 chemical pollutants were found in private wells (Division of Epidemiology and Disease Control, 1980).

3

Lifescape Change: Cognitive Adjustment to Toxic Exposure

Exposure to toxic materials not only changes what people do, it also profoundly affects how they think about themselves, their families, and their world. In short, it represents a fundamental challenge to prior life assumptions. In this chapter, five lifescape shifts are discussed. These involve changes in how victims view their health, their home, the environment, their personal control of the future, and, finally, their ability to trust others.

Although it is nearly impossible for people to insulate themselves totally from the disruption of a toxic incident, the uncertainty of the situation invites a range of interpretations. Generally, residents can preserve their old reality only by denying the implications of exposure. Thus, a Legler resident commented,

> *Today you can't pick up the paper without finding something else that's killing you; we might as well dry up and die. If the water is bad, the hell with it; everything else is bad!*

Similarly, one doubter at Love Canal interpreted declarations of emergency by Governor Carey and President Carter as entirely political (Fowlkes and Miller, 1982). By using a "safe" social explanation, this resident was presumably able to avoid more threatening alternative explanations (e.g., that residents faced a severe health danger).

The prerequisite of most lifescape effects of toxic exposure is that residents believe that they have been affected. Perception, not reality, is the key.

Perceiving a Changed Status

In order to explore the conditions under which residents come to believe in contamination and its possible effects upon them, I will contrast Legler with the Love Canal incident.

Acceptance of Legler's Groundwater Contamination

How did Legler residents respond to notification of water pollution? For most, the news was a serious blow. For many, it was literally unbelievable. And for a few residents, it was even a relief! To understand these differences requires examining the factors that determined whether residents were shocked by notification and those that determined their perceptions of vulnerability to exposure.

Hints of Problems. In the prior chapter, I noted that the announcement of Legler's contamination was foreshadowed by a number of clues that enhanced residents' readiness to believe the warning. Not all residents witnessed such clues, however. Their absence made it initially easier to deny the threat.

> *When we were notified in December, we didn't want to believe it at first. Our water wasn't foul. It tasted good. But when they tested it, it contained acetone and benzene.*

In contrast to the more abstract community problem of landfill security, personal problems, such as loss of water quality and family health, appear to have had more influence on people accepting that their own home's were affected by the pollution. For families that had previously reported detectable deterioration in water quality, the notification confirmed their concerns.

> *I had suspected this for a long time. I wasn't that surprised. It confirmed how I felt. I was angry. I had been complaining for a long time, and the township told me I was crying wolf.*

Furthermore, notification was a sign of hope that finally the township would act; what had been viewed as a private problem was now seen as a communitywide issue.

The report of pollution also provided an explanation for what had previously seemed inexplicable health problems.

> *I made a connection between the water and health problems. In Staten Island we didn't have diarrhea and the water didn't smell.*

For those living outside the first area designated as polluted, the period during which the contamination boundaries were gradually extended outward provided an opportunity to prepare for the bad news. However, the delay in notifying residents more often created false hopes that their homes were safe. Government officials directly encouraged this denial by reassuring residents outside the current pollution boundary that they had no problem. When these hopes were dashed, trust in government was strained.

> *We had received no notification, but I read about the pollution in the* Asbury Park Press. *I called the township and they said that if I didn't get a letter, there was no problem. A person moving onto the street called to ask if the water was polluted, and we said "no." I had been reading in the paper about [route] 571 and Lakehurst. It got closer and closer. I kept thinking about those poor people.*

Perceived Vulnerability. Location was also an indicator of vulnerability, with residents who perceived themselves to be far from the landfill feeling most secure. However, the most important correlates of perceived vulnerability to pollution impacts were the residents' age and the presence of children in the home. Even more than belief in pollution, these factors influenced residents to heed the health advisory and cease using their groundwater. Thus, it appears that residents were not willing to take a chance with their children even when they doubted the veracity of the notification.

> *It took two weeks for us to gradually accept that our water was polluted. We would forget and drink the water during this time. We were more careful with our daughter than we were with ourselves.*

Because nearly all young adults (those under forty) in Legler had children in the home, the parents' age and the presence of children were highly correlated with each other.

In ceasing use of their tap water, these young parents were motivated by a concern about the health effects of toxic exposure.

> *When I heard that the chemicals were carcinogenic, I didn't know what that meant. When it was explained—cancer! My stomach fell. I worried what I had brought my son into.*

The comment of a childless adult provides a point of contrast.

> *We'd be a lot more concerned if we had kids.*

Younger adults also recognized that they might themselves suffer health effects from exposure. As one middle-aged resident noted,

> *I still don't know what will happen in ten years. My husband knows all about Love Canal. I'm very worried about cancer—period! We both have been losing our eye sight; is this related? We don't know what will show up with the chemicals. We were swimming in it!*

In contrast, the few elderly couples in Legler were relatively free of personal health concerns about exposure.

> *What do we have to lose? Cancer takes twenty years to develop.*

However, some worried about the effects on children.

> *We had brought up three grandchildren who spent much time here. Is it possible that this will affect their lives? How can you tell?*

Compared with parental concerns, the length of residence in Legler was a weaker factor influencing assessments about personal and family vulnerability. Thus, although longerterm residents had a particularly strong concern over their children's health, parental concern among newcomers was also evident.

Acceptance in Love Canal

It is instructive to contrast toxic incidents to test the generality of findings. While the general pattern of acceptance at Love Canal was basically similar to that found in Legler, differences in the incidents and in the communities affected the specific dynamics.

Compared with Legler, Love Canal was considerably larger and more urban, with a stronger blue-collar presence. Development had occurred in the 1950s after the cessation of dumping. A substantial proportion of the community was older and had lived in the community for a greater length of time (Fowlkes and Miller, 1982). Whereas Legler residents had sought a rural, suburban life, residents of Love Canal accepted trade-offs by living in an urban industrial environment and, in many cases, by working for the area chemical companies. The latter factor was associated with a slight tendency to minimize the problem.[1] The presence of minority residents who rented their units in Love Canal was a further difference (Levine, 1982).

At Love Canal the key conflict concerned the delineation of safe from unsafe areas. In contrast to the lifestyle impact from the loss of water in Legler, where the problem was "fixed" so that relocation was unnecessary, Love Canal residents faced partial temporary evacuations, the loss of the local school, and an eventual "opportunity" for residential relocation.

A variety of contamination clues preceded the Love Canal incident, including exploding rocks, sudden subsidence of soil, barrels that popped from the earth, and areas where plants either would not grow or underwent accelerated growth. Besides these anomalies, some families had suffered aberrant health problems that later provided an "incentive" for them to believe in the chemical threat (Fowlkes and Miller, 1982, p. 101). Nonbelievers tended to have conventional illnesses, while those who believed in the severity of the contamination had suffered from "a strong pattern of unpredictably recurring, debilitating and diagnostically elusive illnesses" (p. 104).

The degree of concern did not vary with location at Love Canal. However, age was an influential factor, with older residents, highly dependent upon neighborhood stability as they neared retirement, likely to minimize the problem. Age was positively correlated with length of residence; the median length of residence for families who refused to relocate was twenty-four years. As in Legler, younger families were likely to be concerned about their children, a concern that influenced them to have a more conservative estimate of potential risk and, thus, to be more likely to relocate. These families had lived in their homes between eight-and-one-half and ten years, and about half saw themselves as permanently settled in the community (Fowlkes and Miller, 1982; Stone and Levine, 1985).

Overall, decisions about relocation were greatly affected by the extent to which residents believed that the contamination was a real threat (Fowlkes and Miller, 1982). These beliefs, in turn, were related in part to educational levels and particularly to the residents' social values, as suggested by Fowlkes and Miller (1982, p. 96):

> "Non-believers" espouse a highly individualistic and meritocratic set of values. They are defenders of the *status quo*, and subscribe to the view that life in present-day American industrial society is inherently and pervasively risky. Accordingly, they hold that the major burden of responsibility legitimately resides with each family to secure the information and resources necessary to safeguard its own welfare . . . . The "believers" live in less privatized and more sociable worlds. They articulate an inextricable linkage between individual and collective welfare and an expectation that the polity properly stands for the interests of the individual where these would be compromised or jeopardized by the interests of the industrial order.

Believers and nonbelievers differed in their approach to defining the disaster. Believers were more likely to have had uncommon health problems and other direct experiences that they related to chemical exposure. They actively searched for clarification of the situation, attending meetings, questioning officials, and seeking tests. In contrast, nonbelievers had fewer direct impacts from chemicals and relied on their own limited experience and that of close acquaintances. Compared with believers, they were far removed from events and information sources, conditions that reinforced their denial (Fowlkes and Miller, 1982).

Perceiving a Changed Status: Conclusions and Summary

The importance of these indicators of acceptance that contamination has occurred is further confirmed by several studies in New England. Near toxic exposure sites in Acton, Massachusetts, Williamstown, Vermont, and Milford, New Hampshire, Hamilton (1985 a and b) found the greatest concern about contamination among younger and newer residents, the most affluent, and women with children less than eighteen years of age. He stresses the importance of motherhood in highlighting the gap between younger women active in the citizens' groups and the older men often representing government and industry. The importance of parenthood is further confirmed in a study conducted in Williamstown, Vermont, by Ottum and Updegraff (1984).

In summary, prior anomalies help to prepare some residents for believing in the possibility of contamination. Additionally, the vulnerability of a person's family appears to influence belief in the incident and the willingness to employ protective measures.

Once victims of exposure learn of their exposure and come to accept the truth of this information, it is common for them to experience a number of changes in the way that they view life. Five consistent lifescape changes are associated with the acceptance of residential toxic exposure.

1. A reassessment of the assumption of good health.
2. A shift to pessimistic expectations about the future, resulting from victims' perceived loss of control over forces which affect them.
3. A changed perspective on environment; it is now uncertain and potentially harmful.
4. An inversion of the sense of home involving a betrayal of place. What was formerly the bastion of family security is now a place of danger. Having chosen to live there, the person is now deprived of the choice of leaving.

5. A loss of the naive sense of trust and goodwill accorded to others in general; specifically, a lost belief that government acts to protect those in danger.

In the remainder of this chapter, each of these long-term impacts is explored in detail.

Perceptions of Health

One primary impact of toxic exposure is the perception of danger to the health of victims or their families,[2] evoked in this comment by Creen (1984, p. 52): "We kept hearing phrases like 'possible carcinogen' or 'suspected mutagen.' These phrases strike a person like the rattling of a chain—with a sense of dread." Under such circumstances, victims readily become preoccupied with health concerns. Past and current symptoms are attributed to exposure. And given the delayed onset of environmental health problems, expectations about the future are dependent less upon current symptoms than upon perceptions of what victims believe will happen as the result of exposure. Frequently the result is that anxiety about future illness, a shortened life span, and genetic damage cast a shadow over the future. The mystery and uncertainty that surround medical effects may contribute to personality changes as well (Vyner, 1984).

The emergence of health concern in the Legler case study has been described. In this discussion, I will also draw upon observations from other communities suffering from toxic exposure.

Health Preoccupation

When I interviewed a number of Love Canal residents in 1979, they were totally preoccupied with health problems. Despite my repeated attempts to change the subject, the respondents persisted in describing at length not only their own maladies but those affecting their families, friends, and neighbors as well. Unexplained illness dominated their experience. Health problems had been the cause of worry, of financial hardship, of lost work, and of lost loved ones. After being linked to the contamination, these health problems had become an interpretive overlay for life, as illustrated by the response of a resident asked when she had first learned of Love Canal.

> *Love Canal? Last spring. I had no idea whatsoever. I was so tied up with my child being sick, with our own illnesses. And then we went away for vacation, and we came back, and my best friend said, "Hey, the state's going*

to move me out of here!" And she was totally confused. Her husband died from cancer. Her son died from cancer. She had a hysterectomy from cancer— as I did—twenty-nine years old and I had a hysterectomy! I lost a child here.

The list went on and on.

The burden of coping with continuing illness has its own lifestyle and lifescape impacts. Illness focuses the attention of family members inward as resources (financial, emotional, and energetic) are marshalled to deal with health problems. The involved family rarely has the opportunity to question the causes of illness. Seen as a private problem, there is also little attempt to look for patterns of illness in the community. These patterns are rarely visible until contamination is identified and becomes the focus for local communication.

During a period of health concern, the family is dependent upon physicians reluctant to accept environmental explanations for symptoms, relying instead upon conventional explanation and treatment. After being diagnosed as "normal," the illness becomes demystified, and the victims have no reasons to search for environmental causes. Even "unconventional" diseases rarely prod physicians to look for environmental sources.[3] It is no wonder that many victims have the jaded view of the medical establishment held by my Love Canal respondent.

Last summer, I scrubbed my basement. Soon after, I felt real sick. And my hands felt like they were being crushed, and my feet and my spine. And I went to doctors, and I had steroid treatments. And nothing helped, and I went to specialists. I still have it, [it's] affecting my hip now. They don't know what's wrong with me. I don't even want to take my son to the doctor anymore. It's bullshit. They're afraid to open their mouths for fear of losing their license.

When doctors fail to "legitimize" toxic exposure as the cause of health problems, claims by victims may be viewed as irrational (see Fowlkes and Miller, 1982). Furthermore, establishing a rational basis for health impacts is epidemiologically difficult under the best of circumstances.[4] Resulting uncertainty about health effects at Love Canal helped to split the believers from the nonbelievers (Fowlkes and Miller, 1982).

After exposure is announced, families with existing health problems may lack the financial resources needed to take protective action. When I asked my Love Canal informant why she hadn't moved, she replied,

We can't afford to move out. We have everything invested in these homes. We are so swamped with medical bills! My son's allergist said last fall to get the kid out of the house. I'd like to, but I have no place to take him. If I get

another house or rent an apartment, how am I going to keep that up and the house too?

Changed Interpretation of Health

As a lifescape shift, perception of health is altered across all time frames: past, present, and future.

Reinterpreting Past Health Problems. Discovery of toxic contamination provides a framework for explaining previous health problems, particularly those originally inexplicable (see Fowlkes and Miller, 1982). For example, the tragic loss of a Legler child to a rare form of kidney cancer prior to the discovery of contamination was explained definitively (in the view of the family and others in the community) by the later discovery of toxic chemicals in the groundwater. Other examples of the reinterpretation of health problems were suggested by my Love Canal informants.

My son quit growing. He lives in the lower level of a raised ranch. He developed ulcers; he has sugar; three out of five of my kids have ulcers. And it's not because my wife is a lousy cook!

My son had to come home for lunch. And I had to go pick him up and take him back because by the time he'd get to school he'd have an asthma attack. And I used to think it was pollen; he'd get it in the dead of winter!

In response to the health concerns in Legler, the Concerned Citizens Committee conducted a health survey of residents using a form developed at Love Canal. Athough the state health department would later conclude that the survey failed to demonstrate any health patterns beyond possible skin and eye irritation, the survey process heightened awareness of the health issue. Furthermore, the state acknowledged that no conclusive test of health effects could be done with such a small and uncontrolled sample, suggesting to residents that health effects might in fact be occurring. It also pointed to long-term illness such as cancer as "the only plausible health consequence of consuming this water," a conclusion that was hardly reassuring (Division of Epidemiology and Disease Control, 1980, p. 11).

The state report did little to change the link perceived by Legler residents between chemicals known to cause symptoms and the occurrence of those conditions.

When the state described the problems from the chemicals, my husband's kidney problems seemed to fit. Within a thousand feet of here, we have eight people with kidney problems. Do people with a predisposition to kidney problems choose to live near a landfill?

I had blamed my physical problems on getting older. Now I started to think that maybe they were due to the water. I had skin problems, stomachaches, and menstrual disorders. I lived on Maalox. Now whatever goes wrong, I think that it's the water, and it's going to kill me. I'm a hypochondriac.

While Fowlkes and Miller (1982) suggest that unconventional diseases are the most likely to be questioned, my own data suggest that even conventional health problems are reinterpreted after toxic exposure is announced and patterns of ill health come to light in the affected area. Thus, one of my Love Canal informants indicated:

We have several women right around where I live who had hysterectomies. Five men have had open-heart surgery. Almost every child from twelve years on down has allergies, asthma, hyperactivity, above-normal intelligence, or below-normal intelligence; there are very few normal children. On my street alone, one woman moved in next door to me. She had two small children when she came, so she could conceive. She lost five sons. She's 26 years old and had her tubes tied. The woman across the street moved in with one child and lost three. The woman next to her was two weeks away from having her baby when she moved in. The week before she was due, the doctor said everything was okay, "you're beautiful!" Well, the child was born dead.

Reinterpreting Current Symptoms. Reinterpretation of health status also occurs in the present, as this comment by a Legler resident suggests:

It seems silly sometimes when you fear something. For example, I had a tumor in my ankle during the water situation. I thought it was cancer from drinking the chemicals. I expected to find cancer. I was really upset. Even when the doctor said it was benign, I was still worried.

Similarly, at Love Canal, a resident displayed a picture of a girl with chloracne, a dioxin-induced skin condition.

When they started digging, my daughter did not have one blemish—nothing— on her face. And she broke out in the "Hooker bumps," which several other children did in this area.

In addition to reassessing their health, people may question what behaviors are "healthy." In the Michigan PBB contamination case, many mothers grew to fear the consequences for their infants of breast-feeding, creating a dilemma because they otherwise viewed this as a superior means of providing sustenance (see Hatcher, 1982).

Future Health Concerns. There are also changes in long-term health expectations due to toxic exposure. The potential for long-latency health problems clouds the future. Beyond cancers there is the threat of changes that extend into future generations—both mutagenic (changes affecting the genetic material) and teratogenic (fetus-threatening) effects. A Legler woman recalled her concerns as she anticipated the birth of a child:

> *The doctor said that the baby would be aborted if there was a chemical problem. It was in the back of my mind until it was born that it would be deformed. It wasn't until after it was born that I could learn to cope with the water problem.*

Such effects involve more than a sense of violation and intrusion; the seeds of future misery have been planted, and it is only a matter of time before they germinate. As a result, as this comment by a Love Canal resident indicates, formerly happy anticipations may be shifted diametrically:

> *When a child is born in this area, no one says, "was it a male or a female?" They say, "was it normal?" That's the big thing, "Was it normal?"*

Such concerns are not spared children. One Love Canal mother made the following statement while her son stood at her side.

> *I'm thinking about my son. What kind of kids is he going to have if he lives— he has to have a blood test twice a year for leukemia—are they going to be mental retards?*

A Legler teenager reflected similar concerns.

> *What worried me is my genes. This stays with you. It may not show up for twenty years. It may hit the next generation. Having a deformed child can be an emotional crisis. I think of the mentally retarded class at school. I don't know how to handle this situation.*

Ironically, government programs aimed at ongoing health screening for toxic victims may serve as a continuing reminder of the potential for future disaster, maintaining this lifescape shift. Testing may also serve to dehumanize victims, as one Love Canal resident intimated.

> *We're human beings, we're not guinea pigs. We've had blood. We've had all kinds of tests. They treat us like we're a better strain of white mice.*

Toxic victims who accept the full "worst scenario" of health outcomes are haunted by the possibilities, suggesting the palliative virtues of denying such impacts. Accepting the reality of health impacts from living in the Love Canal neighborhood was an emotional burden that for some became insurmountable.

I have a friend who lives two doors down from me. And she just got out of Memorial Hospital for the fourth time in the last six months. She's suicidal. She's classified as schizophrenic. She lost three babies here. She used to hold a secretarial job with a prominent firm. Now she can't function. We're in the process of getting her into a home. Her family can't handle anything anymore. I have another friend who committed suicide. Why? We don't know. The nervous system just can't tolerate it. I myself have been under extreme depression—no reason. It just comes on you and you don't know how to handle it.

Aside from the possible somatopsychic effects of chemicals to which these women were exposed, a mental health impact might be expected in anyone forced to address such all-encompassing health concerns, as Campbell (1981, p. 210) reports. "Poor health appears to have a peculiarly insistent ability to reduce one's sense of well-being, an ability which most people find impossible to resist. Not very many people say they are dissatisfied with their health, about one person in ten, but those who do show an impressive pattern of ill-being—not very happy, dissatisfied with life, and high feelings of strain."

Perceived loss of health in the face of toxic exposure normally carries an extra piece of interpretive baggage, as Creen (1984, p. 53) notes. "Lung cancer to a heavy smoker [is] accepted [as] a person's choice to smoke. Miscarriage to a young mother is not so easily accepted, particularly when there are existing fears for the safety of one's drinking water."

The thought that loss of health was caused by another human's actions is generally disheartening because the loss appears to have been avoidable. However, it can also inspire the fortitude to turn health concern into activism, as this quote from a mother at Love Canal indicates:

We did not ask for Love Canal. It came into our homes, our yards, our vegetables and it's a principle now because we want to fight for everybody in this country.

Environment

What happens when people come to define the environment as being a perpetual source of threat? Is it possible to have a lifescape shift toward an ongoing view of environment as a hazard? Legler's trans-formation from a suburban idyll to a neighborhood threatened by its environment has already been described. Here I make a more general examination of this lifescape shift.

Historically there have been periodic shifts in perception of the environment (Kameron, 1975). A little more than a century ago, hatred of the wilderness characterized Western culture. Conquering the wil-derness and taming it with civilization were the goals of American settlers moving west. With the wilderness since vanquished and reduced to scattered controlled parklands, a shift in the place of danger emerged: "Ironically, the wilderness experience is now seen as being enriching and enlightening, while the city, which was formerly associated with virtue, is now often seen as the bestial jungle (Kameron, 1975, p. 3)."

Ruining the Suburban Myth

The move to the suburbs and quasi-rural areas, undertaken by many of the toxic victims discussed in this volume, was part of an escape from the city to a rural idyll. Likewise, America was seen by settlers over the past two centuries as a new Garden of Eden that could cleanse and purify the blight of industrial pollution experienced in Europe. As a result, Americans failed to anticipate the conflict inherent between the image of America as a bucolic paradise and unchecked industrialism (Kameron, 1975; Marx, 1964).

This conflict is readily apparent at Love Canal near a "Scenic Drive" that juxtaposes massive chemical plants and the roaring Niagara River. When I visited the canal in 1979, huge earth movers were in the process of making a clay cap atop the site. During the workers' lunch break, one mother and her son led me onto the canal around back of a boarded-up brick house, formerly home to their closest friends. When the mother said of the barren landscape, "This used to be a yard," her son interjected, "The most beautifulest yard that you could ever think of." She continued:

> *Tony, who died, spent many years filling in top soil in this yard. And it kept sinking, and they didn't know where it was going. Right here there was a cherry tree and a pear tree, and we canned. And she had a garden right here. And she had huge tomatoes, and we used to say why were the tomatoes and zucchini so big! And we found out that some of this stuff [the chemicals] accelerates growth and some of this stuff kills. There were spots where the grass wouldn't grow.*

The contamination of the garden represents a shift full circle. Now there is no place to escape. This can be seen even in the most remote and bucolic of settings.

Contaminating the Bucolic Retreat

For generations, the small communities along the Tennessee River in northern Alabama had lived in close harmony with the water.[5] Around the small town of Triana, the largely black population had evolved a local culture centered around the river and its tributaries. The major sources of protein were catfish and other aquatic creatures. Fishing was more than a source of subsistence and income—it was a way of life. Fish fries were important family and community events. Swimming and fishing outings occupied vacations and weekends. People spent a major portion of their time with friends and relatives at their favorite fishing spots. They returned again and again to these locations, learning the habits of the fish in each locale.

With the growing recognition of the dangers of the pesticide DDT in the 1960s, attention had gradually turned to the possible impacts of discharges from a major DDT manufacturing site located upriver from Triana. Although domestic use of DDT was banned in 1971, the presence of the compound continues throughout the region. As with other complex synthetic substances, DDT does not break down in water. Instead it persists in the environment, bioaccumulating in the fatty tissues of animals up through the food chain. DDT was found to have significantly contaminated much of the wildlife around Triana, including the fish.

It was not until the late 1970s that the implications of this contamination dawned on people. A study by the Centers for Disease Control (CDC) found evidence of DDT in the blood of local residents. Warning signs were posted along the river. A series of lawsuits sought damages from the manufacturer and others. As residents they heard about the contamination on their televisions or at local gatherings, they were faced with a choice analogous to that confronting other toxic victims. They could continue eating local fish or they could stop. The choice had major financial and social implications. But it altered the lifescape in still another way by changing the perception and role of the local environment in the lives of residents. For example, one resident told me:

> The river—you don't dare touch it with your hand. It's poison. I remember the pleasure and enjoyment. Now it's a dirty place and I wish it would go away. Now I don't go that way.

The Environment as Malevolent Force

When toxic exposure occurs, environment, as the carrier of the contamination, is itself reassessed. Thus, the environment becomes a much more significant, and ominous, component of one's world. Such requisites as air, water, and soil, normally assumed to be freely available in desired purity, are no longer trusted to be safe. Children learn to ask of hosts, "Is the water safe to drink?" I have noted the same phenomenon at every toxic site that I have visited. Some aspect of the environment, formerly benign, is now defined as a threat.

This emergence of the environment as a major force in our lives is a significant shift, exposing an inherent flaw in our worldview. With the transition to industrial society, environment came to be perceived as an object rather than as a surround (or ambient). As a result, we tend to view the environment as a resource to exploit while overlooking the partnership with nature that we belong to as living organisms.[6] Insensitive to our interconnectedness, we act with cultural blinders on. And in the absence of feedback from the environment, there are shocks (see Botkin et al., 1979), illustrated by the famous parable told by the Gestalt psychologist Kurt Koffka:

> On a winter evening amidst a driving snowstorm a man on horseback arrived at an inn, happy to have reached a shelter after hours of riding over the winter-swept plain on which the blanket of snow had covered all paths and landmarks. The landlord who came to the door viewed the stranger with surprise and asked him whence he came. The man pointed in the direction straight away from the inn, whereupon the landlord, in a tone of awe and wonder, said: "Do you know that you have ridden across the Lake of Constance?" At which the rider dropped stone dead at his feet (Koffka, 1935, pp. 27–28).

Toxic victims likewise have trusted a seemingly safe environment only to discover that it was treacherous. Through their shock, they gain a strong sense of the interrelatedness and vulnerability of natural systems. There is a loss of the sense of protection formerly assumed to be present in the environment (see Wolfenstein, 1957). Environment becomes much more important to their understanding of life than it was previously likely to be. Victims may lose their belief in dominion over earth that characterizes the view of Western civilization. Ironically, it is this belief, false as it is, that is a key basis for the individual's sense of personal control. It is therefore not surprising that, beyond a reassessment of environment, there is also a parallel shift in the understanding of self and the future.

Loss of Personal Control

The social psychology of Western society has at its core the postulate that people need to understand, feel in control of, and be effective in producing changes in their physical and social environment.[7] Threatening events can shatter the victim's basic assumptions about the world, giving way to new perceptions marked by threat, danger, insecurity, and self-questioning (Janoff-Bulman and Frieze, 1983). In the wake of toxic exposure, victims not only lose their sense of control, but corollary challenges confront many other of their most cherished personal beliefs.

In Legler, the issue of lost control was a dominant theme for residents (see also M. Gibbs, 1982). When I asked Legler respondents to contrast their sense of control before and after the water pollution, only about a fifth indicated feeling in control at both times. The largest group of residents reported going from a previous sense of control to a current loss of control, as suggested by this comment.

> *I tell myself that I'm in control, but I don't feel it. I feel like I'm in the Twilight Zone. Simple things are out of control. You get to a point where you don't know whether you're coming or going.*

Virtually every element of the Legler situation except the grass roots response robbed residents of control. Their predicament was human-caused; others acted to disrupt their lives. It was an involuntary situation. Management of the threat was controlled by outside forces. The residents' dependence on delivered water resulted in small hassles that reinforced their feelings of helplessness. Legler residents also lost control over their private wells. They were now forced into an expensive central water system, further limiting their freedom.

> *I had always thought I was in control. I own my own home, but I have no say over it. I've been hit financially. Now I can't water the grass with the old well, I have to pay for the new water.*

Beyond the mundane, their loss of control also came home to people in contemplating the significant. Neither their physical nor social environment appeared to deserve continued trust. And most disturbing of all, victims feared that their ability to secure a healthy future for their families was now compromised, as this Legler resident suggested:

> *I'm never going to go into another house with the naive enthusiasm that I had here. I'll be wary and distrusting. I hope the kids don't get sick. I don't believe it though; I think they will get sick.*

A key element in the loss of control was the inherent uncertainty of the situation.

There is no sense of certainty. People have plans and they're shot to hell. I thought I'd live here the rest of my life!

Even Legler residents who claimed to maintain a sense of control felt it was usually diminished in some way.

I'd say yes [that I'm still in control], but I don't think that I'm as smart as I did because I wouldn't have got myself into this situation to begin with.

Others saw the loss of control as temporary.

During the crisis we lost our control. We couldn't do anything about the problem. We talked. We fought over what to do. It didn't make any difference. We had control when we bought the house; afterwards we had no control. We'll return to feeling in control when this is solved or the lawsuit is over and we know where we stand.

But in most cases, Legler residents no longer felt secure.

I'm always under tension. This has disrupted my peace. I have anxiety. I worry about the future. I'm down on this piece of property. I'm at the mercy of forces over which I have no control.

I felt secure, but not now. I don't know what will happen. I live in fear. What will come next? I'm a confused person.

Exposed to unpredicted events beyond their control, Legler residents had lost their illusion of invulnerability. No longer could they claim "It won't happen to me." Not only had they failed to protect their families before, but there was little to suggest that they could be protective in the future (Janis, 1971). The sense of immunity characteristic of our culture (Wolfenstein, 1957) helps to explain the otherwise seemingly careless behavior of homeowners who either failed to seek out or else ignored clues that might have suggested future danger. Yet after a disaster, the immunity is gone. This point is elaborated by Wolfenstein (1957, p. 153; see also Janoff-Bulman and Frieze, 1983).

It would seem for a disaster victim that the world has been transformed from the secure one in which he believed such things could not happen to one where catastrophe becomes the regular order. In his drastically altered view a catastrophic universe has come into being. His underlying feeling may be that the powers that rule the world have turned against him, have declared their intentions to get him, and, if he has escaped this time, they will try again.

Another related outcome of victimization is the loss of a meaningful world (Janoff-Bulman and Frieze, 1983). People are likely to question why they were struck by undeserved disaster. Victimization contradicts their "belief in a just world" (Lerner, 1980), one of the assumptions made by people in order to have the confidence to live. This belief suggests that people deserve what happens to them. As a result, victims are likely to find others blaming them for what occurred. Victims may also blame themselves, sensing either that they were somehow to blame for their circumstances or else that they are no longer in control of events which confront them (see also Peterson and Seligman, 1983).

For some victims, one consequence of this uncertainty, insecurity, and vulnerability is a loss of the ability to plan (see Peterson and Seligman, 1983). Thus, Legler residents commented:

> *I've switched between being in control and fate. I feel that I make a few decisions but that circumstances control.*

> *Before, I had great expectations. I looked for better things. Now I'm afraid to make a move because I don't know how things will result.*

The sense of future among Legler residents was affected by their intentions either to stay or move when they became able to make a choice. Those planning to leave tended to think of reformulating their lifestyles through the move. They spoke of creating a simpler and better life for their families and of escaping government interference. As one resident commented,

> *I never want to have to depend on the state for happiness. We need self-sufficiency. We don't want police, taxes, and a disinterested government. I'll decide what I want to do with my own life. That's what I thought I had here.*

Some residents revealed little concept of the future at all, speaking of living in the short run because of the uncertainty involved in thinking too far ahead. In fact, if there was one projection uniting residents who planned to stay with those who wanted to leave, it was a recognition that wherever one goes, there is no escape from the worry associated with the past exposure to toxic chemicals. All share a changed lifescape regarding health. As a result, their image of the future is clouded, as bluntly indicated by one of my Love Canal informants.[8]

> *If you stay here, you're going to die. And then they say that they don't know what you died from. 'Cause cancer looks like cancer. If anybody's lived here eight or ten years, they will develop cancer.*

The Inversion of Home

The meaning of home as haven from a complex society is inverted by toxic exposure. The psychological importance of this lifescape shift can best be understood in light of the significance of home in American society. "Home" is more than a referent to a structure. It connotes a private place separate from the public that helps to center our lives (Hayword, 1976). It is a place for relating to intimates. Home serves as a basis for two key psychological factors, security and identity. The home permits us to separate and defend ourselves and our possessions from external threat. We further assume that the home itself will not harm us (Goffman, 1971). As a result, we feel secure.

Beyond providing us with a defensible boundary, home serves as the basis for anchoring our sense of self. Houses encode a variety of messages that may be seen as reflective of the owners or occupants in various ways (Ruesch and Kees, 1956), affecting both how we are viewed by others and how we view ourselves (Edelstein, 1973). Much as the facade is a means of impressing others, the interior of the house may come to express the self-concept of the occupants (Cooper, 1971). The house is thus simultaneously supportive of self and revealing of the nature of self.

Cultural identity is also expressed by the home. The "American dream" centers on the nuclear family (Altman and Chemers, 1980) and ownership of a home surrounded by trees on an acre of land (Becker, 1977). Single-family houses express the American values of independence and individuality (Altman and Chemers, 1980), managing social contact by fences rather than by norms. And by separating where one lives from one's place of work, the suburban home further symbolizes both the ability to differentiate one's private life from one's work life and to distance oneself from others, including the extended family. Furthermore, in the United States, the ownership of a house signifies a family's achievement of a desired developmental status, as well as what might be called the economic "creditability" necessary to obtain a mortgage loan. The status results from being found worthy of indebtedness (Perin, 1977). The result is "Not being a 'nation of shopkeepers,' America is one of homeowners, busily investing in plant maintenance and expansion with both money and time, keeping the product attractive for both use and sale" (Perin, 1977, p. 129). This focus upon possible resale makes the homeowner a producer, not just a consumer. It follows, from this perspective, that as "small-scale traders," homeowners are on guard against any threats to the value of their properties (Perin, 1977).

Reinforcing concern with the economic exchange value of the home is the American propensity for mobility (Rapoport, 1969). Not all

Americans are mobile, however (Altman and Chemers, 1980). Some resist moving even in the face of hazards, making an apparent trade-off between the advantages of their living place and the recognized hazards (Preston et al., 1983). In contrast, the process of involuntary relocation due to public works projects forces some people to move who do not wish to, with a resulting grief response over the loss of home (Fried, 1963). Disasters are another cause of forced relocation (Erikson, 1976).

The Meaning of Home in Legler

Residents in Legler were primarily young families with young children. People owned homes that were not "starter homes" but rather represented the achievement of their residential ideals. The home was the central locus of their activity. For many, home was synonymous with recreation. For a working person, a typical evening might include gardening, house chores, playing with children, some socializing with neighbors, or collapsing before the television. For housewives in the daytime, the focus was on chores, children, and some socializing. On weekends, a substantial proportion of residents divided their time between housework and either visiting or hosting relatives or friends. For a few residents, weekends included attending church activities, visiting social clubs or the nearby beach, going camping, or engaging in hobbies or small-scale farming. Overall, Legler was seen as an ideal place to center one's home life.

The psychological significance to residents is suggested by an analysis of the respondents' answers when asked about what home meant to them. The surprising range of answers reflects a varied image of home within the sample (see also Hayword, 1977). Thus, for some Legler residents, home was a place where they were in control, here no one could tell them what to do. It was a place for independence.

> You can go about your business in your own private way—you can have a dog, hang a picture.

Residents also saw home as a place of security and permanence where they were not at the mercy of landlords. Some saw the home as a refuge, a place to relax, be themselves, and escape from the pressures of life.

Some residents thought of home as the orienting point for their scattered lives; it was the place they came back to. It was a place where they could feel "at home," that they could get used to and arrange so they felt comfortable. It was a place for enjoyment and to entertain friends.

Our family is built around our home. We stay here when we have time; we don't take vacations; this is it. We have a pool. We stay home and enjoy what we have. We party here in our and our neighbors' backyards. We avoid a babysitter this way because we can go in and check on the kids.

For some Legler residents, home was a place for solitude, tranquility, and seclusion. It was a repository of memories—of what they put into it and of their lives as a whole.

Home is my life—hard work, heartache, happiness, love and tragedy [picking up a picture of a dead child]; my life is my home—both the good and the bad memories.

Home was a place for observing changes over time and for attaining a sense of achievment.

I enjoy seeing the plants grow, to see changes. It's like taming the wilderness. We've accomplished so much in ten years. There was nothing here when we moved in. We are proud that we did the work ourselves.

Home was, for many residents, a place to personalize, for connection, and expression.

I built this home, and I'm going to die in it. That's how much I love this home! I built a home, not a house. I built this home from the bottom of my heart; I built everything here. My heart and soul are in it.

Home was a place for raising children and gathering family.

Home is a place to have togetherness within the family circle, a place to enjoy holidays with friends and family.

At the same time, home offered the possibility for avoiding crowding.

It's nice to have the space so that the kids can each have their own room; we like having privacy.

Finally, home connoted ownership, responsibility and investment.

Accentuating their significance was the fact that Legler homes often where custom built. They were thus more likely to express the individual taste and desire of the owner than does the average house. People participated in the creation of their homes and thus were likely to be "invested" in the structures themselves.

What emerges as significant in this examination is the tremendous importance of home in the matrix of people's activities. A new home and an acre of land—these were the lucky Americans who had achieved the "American dream" in Legler.

The Inversion of Home in Legler

Inversion of home involves the negation of the hopes, dreams and expectations that surround the institution of home in American society. In Legler, a "home-centered" repertoire of activities was converted from a primary source of pleasure to a cause for dread.

One indication of the inversion of home was the expressed inability of most residents to recapture their ideal life in Legler. People cited as reasons for their discomfort a distrust of the piped water and fears that the landfill would reopen or that the soil was contaminated by chemicals. This inability to feel that things are back to normal involves more than fears about continued exposure. It also reflects something as subtle as a changed feeling about the place.

> *Just the experience that we had in the beginning, the emotion—it will never be erased. It will always hinder us being able to settle here and raise a family the way we want to. I know that this is my home, but I don't feel as comfortable as if this never happened. I want to feel the way I feel about my parents' home, the way they feel about their home. I wanted these type of feelings; I can't have them here.*

Home is no longer the secure place that it was for Legler residents. They were effectively trapped in their homes even as they saw them as places of danger. What is suggested by this change in the meaning of home is its virtual inversion as a naive concept. The process of inversion of home appears to have two related foundations, psychological and financial. Each will be explored in turn.

Home as a Psychological Refuge. Rather than a place to escape to, with the contamination home had become a place that residents could not escape from. Parents particularly feared the consequences of continued residence for themselves and their children. Thus, home was inverted in the sense that it now was accompanied by a strong sense of fear and insecurity. Rather than buffering the family from the dangers of the outside world, home embodied these dangers. Lingering fears about possible alternative routes of exposure other than through groundwater suggested to some residents that they could no longer regain a sense of security there.

Inversion is also revealed in the way the home was used as an expression of identity. Personalization and even routine maintenance halted as people became reluctant to invest more time or money in homes that they saw as valueless. With the costs of the new water system, others could ill afford home improvements. Many appeared to lose their motivation for their homes; they didn't feel like doing work that just a short time before had been a major focus of their lifestyle. For many residents, this loss of motivation continued even after the central water system was provided.

> *We went to Florida. We hoped the house would burn down while we were away. We had a lot of plans for the house when we moved in. But after the water situation, we didn't paint, landscape, carpet—we didn't do anything. There is no joy in this house at all. I hate the floors, the walls. But I'm not going to fix it up. We won't get our money out of it.*

> *There was a time when I stopped cleaning the house. What difference did it make? I couldn't invite anyone over. No one would want to come. I was ashamed and embarrassed. I'm usually very organized. But not then.*

Of course, there were exceptions to this lack of attention to home. One woman reported that during the period of water delivery she made compensatory efforts to have her house neat and clean.

> *I'm a Dutch girl. I kept working on the house all along. I kept washing and scrubbing to get the smell out. I tried to be extra clean.*

With the provision of another water source, a new phase of the Legler incident evolved. It would appear that residents had the opportunity to recreate their previous lifestyles. But having lost their sense of security in their homes, could these families regain this feeling? When my Legler interviews occurred, some two and one-half years after the pollution incident began, about half the families in the sample indicated a desire to leave, although few had actually tried to move.

With the completion of the central water system, some residents expressed rekindled satisfaction with their homes and neighborhood. These tended to be more recent arrivals who viewed the pollution as a technological problem that was now fixed. They felt that few other places could offer them the same quality home at the same price. They looked forward to continued development in the area. They tended to be more comfortable with a central water system under government management than with private wells. And they shared the belief that pollution problems, already confronted in Legler, might occur elsewhere. For these residents, a return to a privatized lifestyle in Legler was eagerly sought.

Why should we move somewhere else, all towns are alike? We hope they learned a lesson here. We would only move if we were forced to. We worked too hard to get this house in the first place.

While some residents were remotivated to work on their homes, many showed signs of ambivalence.

We accomplished everything but the yard, and I still plan to do that. I figure, if we're stuck with the place, we might as well give it our best shot.

This was our first house. No matter what, we do not want to lose it. We'll rectify the problems. We'll stay here.

I hate it here, but I'm making the best of it. I keep the house up so we'll be able to leave.

Some residents sought to stay less out of enthusiasm for Legler than out of a dread of encountering new horrors elsewhere.

We didn't consider moving. If we were looking for a place, we would not know where to go. The problem is everywhere. If it's not smog, then it's the water. No matter where you move, how do you know you're not moving into the same nightmare elsewhere?

Even the view that the water problem was fixed did not erase concerns.

If we moved, it would not eliminate the problem that we had. It would still not erase our worry about the kids. The water problem here is fixed. It's like the house had breast cancer; it was cut off, and now it's okay.

Residents frustrated in their desire to leave Legler tended to doubt that it would ever regain normalcy.

I would hope to get everyone out of here; make it into an industrial park. No one should live here.

Home as an Investment. It appeared to be difficult for Legler residents to separate the pride they felt in their homes from the question of economic value. This concern reflects both the cultural significance of home and the feeling that residents had been misled in many cases when they invested in Legler. The landfill and the barren "2001" were a far cry from the parks and golf courses that residents had been led to expect. Additionally, residents had been confronted with unexpected costs, including water hookup fees, water bills, and other expenses during the period of water delivery. They feared loss of their certificates of occupancy. The direct threat to the value of home as an investment was magnified by the situation of most of the young families, who had just undertaken a major debt to buy their homes while bearing the expense of building a family.

While a few home sales occurred during the incident, residents uniformly perceived a loss of property value and potential for sale. Several residents reported that they had lost opportunities to sell their homes or that real estate agents had advised them not even to try. One resident recalled, "If you tried to list, they'd laugh at you." Even as they faced difficulty selling their own homes, it became harder for Legler residents to relocate. The cost of homes elsewhere was soaring, mortgages were hard to get, and interest rates were substantially higher than those paid on the Legler property, resulting in the sense of financial entrapment felt by virtually all the residents. People had invested everything they had in their homes, but had lost the option to capitalize on their investment. Also lost was the option to leave.

We want to move, but we haven't tried. The only place we could go is to an apartment, and my husband won't do that. We can't sell; we can't rent another house. So we have to stay here until everything is rectified.

I turned down the job in L.A. partially because the realtor said I'd never get what my house was worth because of mortgage rates, and because the Asbury Park Press *was running ads for real estate in Jackson that had "not Legler" across the top. This may be our last house; am I going to lose my butt on it?*

Beyond such stigmatizing advertisements, real estate agents often aggravated already high tension by reminding residents that it was community-generated publicity that had given Legler its bad name. This was truly a dilemma; without publicity, residents had no means of pressuring government for assistance.

Real estate stigma appears to have affected the entire township, as evidenced by demographic data for Jackson Township as a whole (Ocean County, 1981). Jackson Township shared the generally fast growth rate experienced by Ocean County during the decade between 1970 and 1980, going from 4,804 households in 1970 to 6,514 in 1978. From 1977 to 1979, Jackson led Ocean County in its growth rate, but then the Jackson rate tapered off sharply. Similarly, records suggest a rapid growth in the number of Jackson subdivisions between 1975 and 1978, before a dramatic decline beginning in 1979. While other factors may also have influenced this trend, Ocean County as a whole continued to grow at a time when Jackson Township, widely identified with a pollution problem, abruptly ceased its growth.

To gain compensation for the losses associated with the Legler incident, a law suit was filed by the Concerned Citizens Committee. The action was seen by residents as a means for achieving a sense of finality to the period of disruption. If successful, the lawsuit was viewed as affording some residents the opportunity to leave, even if their houses did not bring full value. It would also allow them to address other lifescape issues, such as health concerns, loss of control, and the desire for retribution against local government. In 1987, the New Jersey Supreme Court decided an appeal stemming from the case in favor of the residents. It remains to be seen what residents will actually do when they receive their awards.

What the Legler case study suggests is that inversion of home involves a loss of the security that people normally feel in owning a home. Home is no longer a place of either psychological or financial refuge. In fact, the interaction of these two types of security itself represented a further dilemma. While many Legler residents saw staying as being harmful to their family's health, to leave was to commit financial suicide. They were damned with either choice.

Generalizing to Other Settings

While the basic process of inversion of home appears to be a consistent outcome of toxic contamination, different facets of this lifescape change are apparent in other cases.

At both Love Canal (Fowlkes and Miller, 1982; Levine, 1982) and Times Beach (Reko, 1984), the first instances of government-sponsored relocation in the United States, some residents forsook relocation to remain behind. Presumably, these families were able to retain or regain their sense of home. Some may have been too attached to the home to give it up. Others may have seen few other options for themselves and felt that they could make the best of it by staying in their homes. At Love Canal, families who remained tended to be nonbelievers in the scope and severity of the problem whose privatized lifestyle allowed them to overlook the loss of community and the abandonment about them (Fowlkes and Miller, 1982).

Reko (1984) documents the stress associated with participating in the federal/state buyout at Times Beach. Aside from complex bureaucratic procedures and delays that seemed inequitable and unfair, Times Beach residents were confronted with assessments of their properties that made judgments about homes that had until recently meant everything to them. Reko (1984) describes one family's confrontation with the assessment company. "They knew they would be talking about money, but what they felt was that ten years of work and effort would now be evaluated. To the company, the buyout offer was a measure of currency; to Art and Karen it was a measure of worth" (p. 41).

Times Beach appears to have had the tightest community structure of any of the contaminated neighborhoods that have been documented (Reko, 1984). The loss of community in the face of the evacuation led one resident to comment, "We are a community without a community. We are lost in so many ways" (Reko, 1984, p. 31). This loss of community, reminiscent of Erikson's description of Buffalo Creek, West Virginia, after a tragic flood (Erikson, 1976), was a major dimension of the Times Beach experience that differentiates it from the more privatized communities in Legler and Love Canal. In order to deal with their loss, a group of residents planned a memorial service for Times Beach. Several hundred gathered on a bridge before the town. Reflecting the importance of religion in the community's cohesion, they threw flowers into the river and recited the following prayer (Reko, 1984, pp. 49-50):

We Remember:	Building Our Houses
	Raising Our Children
	Running Our Businesses
	Talking With Our Neighbors
	Having Happy Birthdays
	Going To School
	Playing With Our Pets
	Fishing In The River
	Coming Home Over This Bridge
We Remember:	This Community
We Believe:	That Life Follows Death
	That God's Son Was Born Among Us And Lives With Us Today
	That As Jesus Was Raised From Death, Rise Daily To New Life In Him,
We Hope:	To Remember Our Community
	To Begin A New Life
	To Remember Our Community
	To Begin A New Life
	To Walk With God

If home blended with community in Times Beach, still another toxic exposure case from the eastern seaboard can be used to suggest the merger of home and livelihood.[9] When a neighboring landfill contaminated groundwater beneath a nursery owned by an elderly couple, they were less concerned about their adjacent home than about their beloved plants. They had taken great pride in raising prized ornamentals, particularly azaleas and rhododendrons, without the use of pesticides. The plants not only represented their present and future means of support but were the focus of their daily activities, their relationship, and their friendships with others. While a new central water source served their home after the incident, they could not afford to water their plants from this metered source. The resulting predicament was described by the husband:

> *Dilemma! If dry weather—water; will contaminate—will die. If not water, also will die. What to do? Very lousy feeling. When I see them whither away, I say "give them a shot." Then they die. It's a useless fight!*

The couple's own disappointment and disenchantment with the land in turn discouraged their customers. Stigma was reinforced by the media. As a result of these disclosures, their business declined; their social relationships were also affected. An inversion of livelihood, as well as home, had occurred.

The degree of impact in this case suggests an important rule of thumb for estimating the effects of toxic exposure. Residential toxic exposure is likely to have the greatest impact where the various spheres of life (family, work, relationships, recreation) coincide with the home. The greater the dependency on home—for those working at home, housewives and househusbands, the retired, children, the homebound, and members of clustered extended families—the greater the degree of stress due to a toxic incident. These groups lack the opportunities for escape and ventilation available to others whose life spheres are more spatially diversified. The result is activism in some cases, as has occurred with concerned mothers in affected communities. In other cases, denial results.

Inversion of Home—Conclusion

Beyond the psychological and financial losses, the inversion of home signifies other lifescape changes. Home is the locus of health danger, exposure to a harmful environment, and loss of protective control. Confronted by a confusing and frightening challenge to the safety of their homes, it is not surprising that taxpayers expect assistance from their government officials.

Loss of Trust

A final lifescape shift involves the loss of trust in others generally and in government specifically. This loss of trust by toxic victims is consistent with a more general trend toward loss of trust in government identified in national polls (see Campbell, 1981). However, the degree of distrust found among toxic victims reflects more than a general social trend. It results from a gradual breakdown of the assumption that others, particularly those in government, will aid toxic victims to make their lives once again whole (Levine, 1982). The response of government frequently falls short of the expectations held by natural disaster victims (Barton, 1969). But with toxic exposure, official actions particularly exacerbate the victims' distress. As a result, victims may be stressed as much by their encounters with government as they are by the knowledge of the exposure itself.

This section first explores the loss of trust generally and in local officials, politicians, and state and federal officials, drawing on specific examples from the Legler case study. Then the situational roots of distrust inherent in toxic contamination episodes are explored, indicating why victims are likely to be disappointed by the response of government officials. Finally, the differing expectations of these actors are reviewed.

Distrust in Legler

Legler residents revealed both a general loss of trust in others and a specific disenchantment with government.

The General Loss of Trust. During my interviews with Legler residents, it was evident that a general lifescape change had occurred for them.

> *I view others differently now. I'm suspicious of their motivation. I expect them just to seek personal gain. Before, I used to worry about what others think. I don't consider it anymore because nobody does anything for you unless there's something in return.*

> *I was very naive; I trusted everybody when we bought our house. Now I can't trust anybody—which is terrible, but I've become very cautious and cynical.*

As with any incident that has negative ramifications for an entire community, Legler residents faced a hostile and unsympathetic reaction from some of their fellow residents of Jackson Township. They felt stigmatized and blamed for their predicament, adding to their feelings of distrust.

> *We were the scapegoat for all their money problems because of the cost of delivering water. They gave no support; they looked at us as lepers who caused stigma and publicity. The township committee put us down as publicity hungry.*

When they closed the landfill, people dumped their garbage on our front lawn. A lady chased me down the street with a broom.

At the award ceremony at the high school, the principal spoke about the need for a positive public image for the town. This was directed at Legler. I wonder what they say at Jaycee meetings? It's as though we are blamed for the situation and don't deserve a remedy!

When a park on the other side of town used fill from the landfill, parents wouldn't let their kids play there. And yet they boo Legler people when we complain at meetings. They give us hell. There have been cops there to stop fistfights. We can drink it, but they can't play in it!

Outside confirmation of these views came from a local newspaper that described efforts by a Jackson Township committee to undo the stigma that had arrested growth in the whole town.

This new group may not be able to completely wipe out the media's negative depiction of our town but maybe they'll be able to finally put to rest the north Jersey impression that every damned one of our 6,500 households in our 100 square miles has a water problem (Miller, 1981).

Stigmatized, disillusioned, and distrustful, many Legler residents adopted a vigilant mode of decision making based on caution and deliberation.

We do more investigation. We don't believe what we hear; we don't trust too many people. When we go to buy something, we think twice.

Loss of Trust in Government. The general level of distrust found in Legler was overshadowed by the loss of trust in government experienced by residents. Jackson Township was viewed both as the polluter and the expected agency of remediation. Accordingly, it was seen as the most deserving recipient for blame, as this comment suggests:

If my children get sick, I'm going to hold each individual on the township board responsible for it. I have this fermenting, boiling emotion.

The lawsuit against Jackson expressed this blaming. Residents felt that they had been wronged, that a moral injustice had occurred. The lawsuit was a means of rectifying the injustice.

I want justice—revenge! This was criminal negligence. Watergate has nothing on Jackson.

Furthermore, Jackson Township officials were seen as decidedly un-helpful during the Legler water crisis. The fact that the township was the polluter as well as the source of government assistance had political, legal, and financial ramifications. The township's resulting difficulty in accepting responsibility for the situation appears to have affected how officials responded to the Legler residents.

> *They didn't believe that the water was polluted even when their own staff said it. They believed that the state had gone wild. They couldn't believe that they caused the pollution.*

While newer residents were particularly angry that the township had issued certificates of occupancy permitting the builders to sell them their houses, all residents were angered by the townships's unsympathetic response.

> *The township board totally humiliated us. For example, one time when I had waited a long time to speak, a board member remarked to me as I took the floor, "Isn't it past your bedtime, little girl?"*
>
> *When we appealed for help, they badgered us and made us into the bad guys. It was like dropping a bomb on us and then making us feel responsible. They even swore at me.*

With few exceptions, the local government was generally viewed as a major part of the problem in Legler. Government served to complicate the residents' attempts to cope with their disturbing situation. As a source of information during the crisis, government officials were seen as evasive, saying only what was expedient, and distorting information that they grudgingly shared.

The township was viewed as having violated its own promises in the way it responded to the situation. While the local officials may have been sincere in their early estimation of their ability to deal with the problem, the fact that they were wrong in these estimates led to a questioning of their sincerity as well.

> *We trusted the town to be adult and to take care of the problem. The mayor promised that it would be fixed in six weeks: "By the time you empty the first barrel of water, the problem will be taken care of."*

Residents were also angry at the way the township responded to their attempts to organize and work on their problems. Some residents charged that the township used threats to try to quiet residents. One active Legler resident was allegedly fired from a township job. Officials were accused of deliberately trying to cause tension within the ranks of the community group.

Much as did residents of Love Canal (Levine, 1982) and Times Beach (Reko, 1984), Legler residents spoke about how astonished they were that their status as taxpayers meant so little. Many felt that the township sought to avoid the problem or even deliberately covered it up. Board members were suspected of playing down the problems before election time. Legler residents voiced a virtually unanimous belief that local government cannot be trusted. Some reported that they would no longer vote. Some of the most bitter disappointment was voiced by young people.

> *It makes you wonder when grown people in authority do this; it makes you wonder where it's all going to.*

Residents born abroad who had adopted this country were also very upset.

> *My husband was a super patriot. Now he refuses to vote.*

Similar disillusionment and loss of trust in local government appear in virtually all the cases mentioned in this volume. Perhaps with the exception of Times Beach (Reko, 1984), local government appears to be subject to so many contrary forces that distrust is likely to occur.

At Love Canal and Times Beach, mistrust was most heavily focused on state and federal agencies.[10] In Legler, where the township was seen as responsible both for causing and solving the pollution, state agencies were seen as comparatively more sympathetic (ICF, 1981). However, it was apparent to Legler residents from the actions of state regulators that they could not rely on their help.

> *At an early meeting we saw different agencies bickering over who was responsible. It was scary. There was rhetoric, but no action. It was frustrating. The people believed that the water was bad; they were hypochondriacs. And there was no one to ease their minds. The government was waiting for a dead body which could be proven to have been caused by the chemicals. One agency after another said we had to prove this. Tests were done, but no one ever came back to warn us about the findings. No one had a sense of responsibility.*

Meanwhile, Legler residents found state politicians to be exploitive, to the extent that they paid any attention at all. Governor Byrne was probably the most disliked for making an appearance to cut a ribbon at Great Adventure amusement park, a few miles away, while ignoring Legler's plight.

This resident's comment sums up the perception of government found in Legler after the contamination incident:

> *I've lost my belief in government. I always thought that in order to say things in public, they had to be true. Now I realize that government exists to pacify people.*

Examining the circumstances of exposure more broadly will help explain this shift in perspective.

The Situation Invites a Loss of Trust

Distrust is in large part the result of a negative dialectic between citizen and government official (see also Chapter 5). Miller (1984) suggests three key situational characteristics that contribute to this dialectic: the role of government in discovering the problem, the inherent uncertainties, and the pervasiveness of exposure.

Discovery as the Starting Point. Based upon his experience working for the Missouri Department of Health at various dioxin sites, Miller (1984) observed that their part in the discovery of localized environmental disasters results in a special role for health and environmental officials in defining that a threat exists. As a result, residents are drawn into encounters with government agencies at every turn in their attempts to deal with the crisis. For officials, Miller (p. 2) observes, "It means that actions which are usually viewed as exclusively technical matters are transformed into public issues which come to be an object of observation, criticism and intervention."

Under public scrutiny, government decisions are no longer cut-and-dried. For example, at Love Canal, residents' understanding of official response was hampered by their unfamiliarity with regulatory procedures and guidelines. While for citizens the distinction between a health "emergency" and a "disaster" had little meaning, for officials it defined their role by determining the allowed response (Fowlkes and Miller, 1982). Their role in discovering the exposure places agencies center stage in the process of clarifying the threat. It is no wonder that officials sometimes are afraid to release news of their discovery due to a "fear of alarming people."

Uncertainty and Localized Environmental Disaster. Local environmental disasters are inherently fraught with uncertainty. The pollution is not easily identified nor are its characteristics easily described. There is likely to be no reliable basis for estimating the consequences of exposure over the long run. Miller (1984) refers to an EPA statement that reflects this uncertainty: "Dioxin in Missouri may present one of the greatest environmental problems in the history of the United States. Conversely, it may not."[11]

The response of some residents to this uncertainty is to embody what Miller (p. 3) refers to as "risk enhancement." Drawing reasonable conclusions from observations that often are misleading, such as an unexplained illness, they conclude that the situation is highly dangerous. While officials may view this reaction as panic and hysteria, residents view officials as deliberately playing down the severity of the problem (Miller, 1984). Stone and Levine (1984) note that the lack of public participation in decisions, combined with a tendency to allay fear rather than share uncertainty, creates a pattern likely to engender distrust.

Miller suggests that the problem for officials is not so much the uncertainty per se, but rather the need to make some assessment that can serve as a basis for a response. For example, defining who in Missouri was highly exposed to dioxin resulted in a range of psychological and financial consequences that exacerbated the stress otherwise inherent in the situation. Similarly, Barr (1981, p. 125), a physician, describes the pressures on medical experts to clarify fears aroused by numerous incidents involving exposure to polyhalogenated biphenyls (such as PCB and PBB) even though so little is known about their health implications.

> In the face of all this ignorance, we are nonetheless asked by mothers whether or not they should breast-feed their babies. There is no simple answer to this question. It is an area in which passions may prevail over admittedly weak science. Over and above questions of toxicity, consideration must be given to the adverse effects of anxiety about breast milk contamination on the mother-infant pair. Denying that a problem may exist can, in many cases, lead to increased anxiety. Reassurance should help reduce anxiety, but how can one say "I don't know" and sound reassuring?

Harris (1984), a county health commissioner in New York State, is considerably less sympathetic in his explanation for why the public has no tolerance for ambiguity (p. 429).

> Used to neatly packaged television dramas like *Quincy*, where all loose ends are tied and all answers are in before the final commercial, the public finds it hard to accept incomplete knowledge and ambiguity. Far easier to believe some newspaper columnist or community leader who offers certainty than a cautious health official who tries to explain precisely what is or is not known. To a frightened and impatient public, health officials' punctilious concern about the thinness of scientific evidence and their

disinclination to draw conclusions from insufficient data are easily mistaken for lack of resolve or abdication of the responsibility to act.

The inherent uncertainty of toxic exposure situations thus has implications both for the government official and for the toxic victim. In some cases, the official's role is rather like that of solving a very complex puzzle. This is seen, for example, in the description by Drotman and colleagues (1983) of the complex detective work necessary to trace PCBs back through the food chain to an industrial accident that had released them. While the official may wish to tackle the puzzle sequentially, the victims press for early clarity and action. The regulatory response is often described as careful science, but there is another interpretation for it, as suggested by Reich (1983, p. 309). "The definition of a problem tends to become frozen in the position of a bureaucratic agency, and thereby to resist change." He further notes that agencies deal with uncertainty by establishing routines that divert them from anticipating problems. As a result, they must continually react to situations that demand their response. In a toxic exposure incident, the impacts of such bureaucratic rigidity are further complicated by the numerous agencies that may become involved, each with its own experience and expectations (Shaw and Milbrath, 1983).

The result is that, having discovered and disclosed the threat, government is unable to provide much further clarification. This affects all residents of a contaminated community, even those engaging in denial. Thus, at Love Canal, Fowlkes and Miller (1982, p. 119) found that "Ironically, whether residents were disposed to remain or anxious to leave, they received no confirmation or reassurance about the hazard or safety of the situation."

The Pervasiveness of Localized Environmental Disaster. Finally, Miller argues that a localized environmental disaster is pervasive. In catastrophic technological or natural disasters the threat follows an observable pattern of appearance via a single route and then disappears. In contrast, localized environmental disaster involves the appearance of a contaminant in an "occult fashion" (1984, p. 4), such as through the migration of leachate from a landfill. By the time of discovery, residents are threatened through more than one route of exposure. The soil, the air, the drinking water, dust, and moisture seeping through basement walls may all carry the pollutant. In the Missouri dioxin cases, residents were hard pressed to avoid exposure because of the pervasiveness of contamination. As at Love Canal,[12] residents interpreted efforts by government officials to discount certain of these exposure routes as attempts to underestimate risk. After relocation, for example, some Times Beach residents feared that their furniture had absorbed dioxin, thus posing a continued hazard of exposure for their families. Officials faced a hopeless task in separating people from contamination because the contamination was literally everywhere (Miller, 1984).

Toxic disaster is a challenge for both regulator and victim. At Love Canal, there was a recognition by many residents that official response was hampered by the newness of this type of disaster and the corresponding lack of organizational experience in dealing with it (Fowlkes and Miller, 1982). And although toxic incidents have their honeymoon period, it ends when officials fail to produce clear solutions (as they seem inevitably to do) and a collision course is set. Miller (1984) suggests that to avoid distrust officials must learn to communicate about complex and uncertain issues in an understandable and timely fashion. Residents in turn must learn to comprehend and interpret such information in a way that allows them to make decisions. Furthermore, officials must share their constraints and seek to understand residents' needs; citizens must learn to effectively organize and to learn what government can and cannot do for them. Even with these precautions, the elements of distrust are inherent in the situation; the dynamics of localized environmental disaster result in controversy that is inherently polarizing.

Responsibility for Toxic Exposure

Another key source of distrust in toxic incidents involves confusion over the assignment of responsibility. Such attributions allow the victim to understand why exposure has occurred and who is at fault. Even more important for their attempts to gain control over the situation is the identification of responsibility for remedying the situation (Brickman et al., 1982). The distinction between these two kinds of responsibility is a crucial one, having major psychological and social ramifications.

It would take a legal treatise to adequately explore the question of responsibility for toxic exposure. In the context of understanding distrust, I focus here on the government official's dilemma—that regardless of the cause of pollution and the problems in addressing it, government is routinely blamed both for failing to prevent the toxic exposure and subsequently for failing to adequately rectify the pollution and all the secondary problems that stem from it.

It's the Victim's Responsibility. A circumstance that particularly invites citizens' anger and distrust is one in which they "get stuck" with the problem. This results from several factors. First, government action often is statutorily limited, as when well pollution is termed the problem of the owner because a well is private property. The resident is likely to be struck by the contradiction involved in the failure to define aquifers— underground water sources—as a common resource requiring community protection when reservoirs—aboveground water sources—receive government protection. Second, government may also lack the interest, expertise, and resources to act. As a result, citizens must often prove that instances of poor health are caused by an environmental source.

A third situation that often shifts the burden to the citizen involves problems that stem from "nonpoint sources." For example, in the Relocated Bayway section of Elizabeth, New Jersey, there are so many potential polluters nearby that it is frequently impossible to prove that any one party is responsible for contamination and equally difficult to think about solutions.

Fourth, the focus of government response is often to look for polluters who can be held responsible for the costs of cleaning up the contamination. This focus is quite different from that of the affected resident, who is concerned with identifying the degree of exposure, clarifying the consequences, and having protective actions taken to stop the hazard and mitigate the damage.

Toxic incidents often are defined at first as a private problem only to be more broadly generalized at a later time. For example, the Michigan PBB herd poisoning was initially found in a single farmer's cattle. For an entire year, he sought an explanation for his herd's illness, but was told by government agencies that they could not study a problem experienced only by one isolated farmer. The farmer persisted in tracking down the cause, eventually helping to bring recognition to a pattern of contamination that had affected not only hundreds of farmers but thousands of consumers (Reich, 1983).

However, even when its scope was understood, Michigan officials refused to assume responsibility for dealing with the PBB problem. Reich (1983, p. 306) describes the predicament that resulted for farmers:

> But state officials decided *not* to condemn any animals and *not* to order disposal, because they did not want to open the possibility for farmers to file suit against the state or for the state to be held financially responsible. The state thus quarantined farms and monitored the disposal operation, while the farmers and the companies decided on their own whether to destroy the animals. Farmers, in turn, felt enormous pressure to dispose of their animals.

Farmers could depend neither upon the polluter nor government to help them. They were left either to privately absorb the costs or to organize to force solutions. Protests resulting from the Michigan PBB case culminated in a march by farmers to the steps of their state capitol where the carcasses of contaminated cows were dumped (Coyer and Schwerin, 1981). While organizing enabled farmers to force a remedy, being forced to solve problems that they did not create was a bitter experience (see Brickman et al., 1982).

Holding the Polluter Responsible. If there is an "ideal" solution to existing toxic contamination, it would hold the polluter responsible for cleanup and compensation. Most federal and state approaches are based upon compliance and enforcement models for achieving precisely this outcome.

However, in practice, the model of polluter responsibility has been fraught with problems. Where a polluter is identified, available laws may not provide an adequate basis for prosecution. Polluters may seek protection in bankruptcy or in legal delays. Evidence linking the polluter to the incident may be inadequate.[13] Government efforts may be dominated by "making a case" against the perpetrator to the detraction of cleaning up the pollution. Government agencies may strike out-of-court deals (called "orders of consent") with the polluter, settling the case without full remedy from the victims' perspective. Time delays are frequent. Even the Superfund model, which attempted to place solutions ahead of blame, revealed minimal success in its first incarnation. A strengthened bill passed late in 1986 may serve to turn the situation around.

Given their corporate resources, it is not surprising that polluters often escape prosecution. But what is surprising is that the polluter often appears to receive less than its share of blame. What are some possible reasons?

Anger and blame may be mollified where the polluter is a major source of local employment. At Love Canal, for example, Hooker Chemical was able to keep a remarkably low profile during the disaster in part because local residents were adapted to and dependent upon the chemical industry.[14] Hooker was also active in trying to minimize public concern over health threats. Interestingly, while efforts by the state health department to allay panic met with public anger, Hooker was not similarly vilified. Paigen (1982) explains that, while the agency was seen as a public servant possessing knowledge about health hazards,

> The community made some allowance for Hooker because the chemicals were buried many years before the chronic toxicity of chemicals was understood and before regulations concerning disposal of toxic waste existed. Hooker Chemical claimed that it had used state-of-the-art technology in burying the waste and that furthermore they had warned the Board of Education not to build a school on the site. The community also understood that the goal of industry is profit and that Hooker was acting in a manner consistent with its goals by using the cheapest method of disposal (pp. 31–32).

In short, the state was responsible for acting for the community's welfare; Hooker was not. The state was supposed to have expert knowledge about the risks from Hooker's chemicals that Hooker was not expected to have. The state should have foreseen problems from the Love Canal disposal that Hooker was excused for not avoiding. The state should have regulated Hooker when the company failed to regulate itself. In the absence of regulation, Hooker was not at fault. While the school board was irresponsible in allowing development of the area, Hooker comparatively escaped such judgments by the residents.

Further light on how corporations get away with acts of "elite deviance," such as polluting, is cast by Vissing's (1984) description of the Dow Chemical Company in Midland, Michigan. Dow was the principal employer in the town. Keeping a high profile, it further established a reputation as a good citizen and earned what has been termed "idiosyncrasy credit" (Hollander, 1958), whereby Dow was held in such high regard that its deviance was overlooked. When widespread dioxin contamination was discovered, Dow actively managed public perception of its responsibility. The corporation stressed the minuteness of the quantities of dioxin found, presented its viewpoint in a brochure called "The Truth about Dioxin," involved its executives in a public relations campaign, and generally maintained that "Midland is a better, healthier place to live because of the corporation" (Vissing, 1984, p. 15). Citing state studies that failed to find evidence of health problems, the city and many residents firmly backed the corporation, overlooking evidence suggesting that Dow had lied previously about dioxin and had secretly edited a key Environmental Protection Agency report.

Holding Government Responsible. Does government deserve to receive more blame than do corporate polluters? Sometimes government shares the blame for the pollution's occurrence because government regulations, lack of monitoring and enforcement, or decisions to permit the polluting operation do, in fact, contribute to the problem. But most often, government is blamed not for causing the problem but for failing to effectively remedy the situation after it is discovered. After discovering and announcing the pollution, government often fails to move the incident toward closure. At times, government's definition of closure varies from that held by residents, as when negotiated settlements with a polluter fail to address community concerns.[15]

It is the realization that they can not depend on government to solve their problems that often spurs the contaminated community's residents to collective action aimed at forcing a solution. When they confront angry citizens, government officials who expect that citizens will be passive and respectful quickly find otherwise. As a result, distrust emerges from the varied attributions made and expectations held by different parties at different points in a toxic disaster. Miscommunication and mistrust ensue because the parties have divergent views of who is responsible for remedying the situation. Citizens may look to government for help when government either lacks the means to provide it or defines the citizen or a polluter as the party responsible for action. Government officials may expect citizens to trust them to deal with the situation, thereby removing victims from active participation or even full understanding of their circumstances.

Conclusion—The Lifescape Impacts of Toxic Exposure

The loss of trust, the inversion of home, a changed perception of one's control over the present and future, a different assessment of the environment, and a decided tendency to hold pessimistic health expectations—these are all indications of a fundamentally altered lifescape. Such changes do not just occur. They are the result of a process of coping whereby the individual, family, institutions, and community attempt to deal with the newly accepted realities of toxic exposure. The dynamics of coping with contamination are examined next.

Notes

1. Fowlkes and Miller, 1982; see also Vissing, 1984; Francis, 1983; Evans and Jacobs, 1981.

2. See also Freudenberg, 1984a; Fowlkes and Miller, 1982; Levine, 1982.

3. Brown (1980) cites several exceptions. For example, the discovery of the Bloody Run site in Niagara Falls was spearheaded by a local doctor.

4. One rare exception was the Harvard research project in Woburn, Massachussets, which successfully linked exposure to solvents and leukemia (DiPerna, 1984).

5. These observations about Triana, Alabama, were gathered while I was consultant to the law firm of Hogan, Smith, Alspaugh, Samples and Pratt, P.C.

6. See von Uexküll, 1984; also Bateson, 1972; Slater, 1974.

7. See Heider, 1958; Kelley, 1972; deCharms, 1968, for examples.

8. Health psychologists confirm that the questioning of peoples' health status causes them to feel more vulnerable. This vulnerability is, in turn, a cause for worry and a loss of unrealistic optimism (see Kulik and Mahler, 1987; and Weinstein, 1982).

9. This case was prepared for Michael Gordon, attorney, in 1983.

10. For Love Canal, see Fowlkes and Miller, 1982; L. Gibbs, 1982a; Levine, 1982; and Shaw and Milbrath, 1983. For Times Beach see Reko, 1984; and Miller, 1984.

11. Miller, (1984), p. 3, quoting from the *St. Louis Post-Dispatch*, November 14, 1983.

12. See Fowlkes and Miller, 1982; L. Gibbs, 1982a; Levine, 1982; and Paigen, 1982.

13. Sometimes, in the absence of firm proof, rumors may lead to unfair assignments of blame. For example, in the Washington Heights section of Wallkill, New York, it took several months for government agencies to accumulate evidence demonstrating a link between an area industry and the tetrachloroethylene found in residential wells. Because tetrachloroethylene is routinely used in the laundry business, many suspected a neighborhood laundry. In fact, this was found not to be the source.

14. See Fowlkes and Miller, 1982; Paigen, 1982; also Francis, 1983.

15. For an instance where a culpable government was able to shift blame effectively to a responsible corporation, see Shrivastava's (1987) description of the Indian government's response to the Bhopal disaster.

4

Individual and Family Impacts

To explore the issue of how people cope with toxic disaster, this chapter will examine individuals, their families, and the social relations and networks that normally support people during life crises. For individuals, the focus will be the elements of personal growth and psychological damage associated with exposure. For families, the focus will be the pressures that pull some families together while forcing others apart. The special issue of the effects upon children will also be explored. Finally, the isolation of toxic victims from their customary support relationships and the development of new sources of support in the affected neighborhood will be analyzed.

Coping with Exposure: Individuals

Residential toxic exposure is highly disruptive for almost everyone involved. Even those who least believe that contamination will affect them must still confront stress resulting directly from the pollution incident. And as previously described, those who accept the full implications of exposure are likely to face even more disruptive effects. What, then, are the common psychological consequences for victims of toxic exposure?

In Legler, a companion study to mine was carried out by Margaret Gibbs (1982) using quantitative clinical measures in order to independently confirm my qualitative data. In her sample of the Legler community, 96 percent of the respondents reported emotional reactions due to the incident. Beyond health worries, they reported (in declining order of mention) feelings of disturbance, anger, depression, family quarrels, mistrust of others, financial worries, feelings of being trapped or helpless, divorce or separation due to the crisis, nervous breakdown, and interpersonal aggression (such as a child assaulted at school) linked with the crisis. Roughly parallel findings are reported for Love Canal as well (Stone and Levine, 1985). These findings suggest that beyond coping with the stressful events in a toxic exposure incident, toxic victims also must deal with the secondary effects of their stress.

Coping with Exposure

How do toxic victims attempt to gain some measure of control over their stress? As in dealing with any stressful event, some people are likely to cope by confronting the problem in an attempt to master it. Others will engage in denial, changing the way that they think about the problem; feelings of stress are reduced, but the problem is unaffected (see Lazarus and Launier, 1978).

For example, more than half of some 100 recent mothers tested in the 1970s for the chemical PPB in their breast milk had conflicting feelings about whether to breast-feed and half felt guilt over endangering their children. Only about one-third of the group attempted to master the situation by changing their behaviors in order to reduce their infants' exposure to the PPB. Their responses included actively searching for alternatives, altering diet, changing where food was purchased, changing the frequency of nursing, switching to bottle-feeding, moving away, and consulting experts about what to do. The remainder engaged in denial, as indicated by a lack of protective actions, forgetfulness about their levels of exposure, and inability to articulate their feelings. Mothers with the highest levels were found to engage in the most extensive denial (Hatcher, 1982).

In feeling their way between mastery and denial, toxic victims confront a situation characterized by extreme ambiguity. Thus, the situation is capable of being construed in varied and contradictory ways. Possibly the most perceptive analysis of the effects of this uncertainty on whether a toxic victim adopts mastery or denial was this comment by a Legler activist:

> *This was a crisis situation with no specified reaction. There was no grief ritual. You don't know what to do. There are divergent emotions and reactions needed to cope. People prefer that this didn't happen. They can't see water pollution; they don't feel bad. They believe it, yet they can't cope, so they rationalize it. Even I have a point where I say "enough, I can't believe any more." When the [neighbor's] child died, I reached my breaking point. I couldn't believe that he died from the water because I couldn't live here with the kids if I believed this. Other people shut off at the beginning. One person got an ulcer, and the next didn't believe that there was anything wrong. My one neighbor was happy with her "coffee klatch" water club because it gave her something to do. We didn't know what we were supposed to be doing! Are we paranoid, hypocritical crazies? Other times, I didn't take it seriously enough.*

Then I called Michael Brown [author of Laying Waste*] and talked to him. He made me into a "basket case."*

Given their participation in the community group, my Legler respondents were not ideally suited for studying denial of the toxic contamination issue. Still, there were indications of palliative behaviors, even among the leaders. Thus, the wife of a "burned out" activist reported that he watched television so excessively as an escape that his behavior had become dysfunctional.

At least among Legler residents participating in the lawsuit, there were mostly indications of active coping. A number of strategies were employed for such coping.

Communication Strategies. Legler residents learned to use subtle communication strategies in managing stress. They would adeptly change the subject to deliver themselves from endless conversations about the groundwater pollution. They would avert others' kidding about their problems by themselves initiating a joke about the situation. Thus, one couple told their friends,

You can go to Mexico or you can come to our house—but whichever, don't drink the water.

Social Escapes. Social activities were used as an escape from problems and a means for relaxation. Although much of the normal social activity of residents was interrupted by the crisis, residents of one neighborhood continued to party and drink beer on a nightly basis. Other residents realized that they needed to escape periodically from Legler in order to cope with living there. One couple joined two clubs and went dancing every weekend. Another would call up a babysitter and take off. Another couple joined a camper club to get away.

We met sympathetic people who didn't know us well enough to kid us—this gave us a lift. There is something about camping. You can be only twenty miles away and be in a different world. We could run the water in the camper.

Some people learned to escape at home.

I'd take a day to rest. If bills were due, I'd put them off. I'd watch TV, read, go for a ride, visit my mother—I'd stop thinking about it, relax, and try to work things out.

The Helper Role. Still another active coping technique involved escaping into the situation. Some people took on the role of helper, coming to grips with their own situation by assisting others in dealing with the crisis. While activism increased the helpers' own burdens in many respects, it allowed them to step outside the role of victim for a while. It was also a means of learning and discovering things that were helpful to them but which they would not have pursued had it not been for their role as helper. Furthermore, many of the helpers were able to rationalize that by using their experience to teach other communities how to avoid similar problems, their suffering was not in vain. In this way, a negative experience was reconstructed in a more positive light.

Keeping A Positive Attitude. As the last technique illustrates, cognitive adjustments were also made in the interest of coping. Some people sought consciously to put problems into their proper perspective and to attend to their immediate lives. As one person commented,

> *I'll keep on planning what I want to do. If a problem comes, I'll deal with it when I can. I'm not going to sit around and worry about day-to-day life. It's always in the back of my mind, but I don't think about the water problem when I have to change a diaper.*

A related cognitive adjustment was the acceptance that circumstances were beyond the person's control.

> *My sister asked what we'd do if one of our kids got a kidney problem. I pray there is no problem, but I can't do anything else. Sitting and worrying won't help.*

Outcomes—Positive and Negative

The stress resulting from toxic exposure is a powerful force for personal change. On one hand, new experiences encourage new competencies. On the other hand, even in the face of positive growth, coping with the stress from exposure may cause serious and long-term psychological damage. Each of these outcomes deserves comment.

Exposure as a Growth Experience

Positive growth can result from the effective and assertive coping that characterizes activists (see also Stone and Levine, 1985). Mastery of the exposure situation requires innovation outside the normal repertoire of behavior and the normal boundaries of thought. Rather than merely acting in the tried-and-true ways necessary to maintain the status quo, the toxic activist adopts an assertive approach to coping that involves making new relationships and finding new solutions (see Botkin et al., 1979). The result is an opportunity for learning and growth, albeit often in the sense of the cliché "learning the hard way." Enhancements in both competence and confidence are simultaneously the prerequisites and benefits of such coping.

While the Legler experience was generally negative, many residents revealed impressive personal growth. A good percentage learned to be assertive in fighting for their interests. The leaders of the Concerned Citizens Committee were particularly effective in guiding the community's coping response. For the leaders and involved members, activism may not have been part of their prior normal lives, but now it was essential for effective coping.

> It was an education. I never knew you could get active in anything. I became outspoken; I was a fighter!

> You have to keep fighting and trying. I can never again let someone else do the work for me.

Along with a shift to mastery came opportunities to gain personal confidence. As people investigated their predicament, they found themselves frequently in situations that were novel for them. Responding to the township committee at a public meeting, for example, they might find that despite a prior fear of public speaking, they could articulately explain their views. Television cameras presented another testing ground. Many residents who faced them came away from such opportunities with a strong sense of self-confidence. Furthermore, finding themselves in a situation where they were forced to express themselves, residents learned that it was legitimate for them to show their anger. Expressing their feelings was now an acceptable alternative to hiding their feelings.

A toxic exposure incident provides a class in the dynamics of modern politics and bureaucracy. In Legler, as people observed the response of the government that they had always trusted, they experienced a loss of naïveté and a dose of cynicism. The result was a less gullible community, capable of critically examining future situations. The summation of these experiences was that residents developed a confidence that if they could pull through the toxic disaster, they could survive almost any crisis.

> It was hard for me to think under pressure. I felt the responsibility of having my first home, my first child, a new situation, and a new area. I had all of this dumped on me at one time. Now when things come up, it's nothing!

Nationally, the most prominent role model for personal growth in the face of toxic adversity is Lois Marie Gibbs, president of the Citizens Clearinghouse for Hazardous Wastes, whose transition from housewife to the savvy leader of the Love Canal Homeowners Association has been chronicled in a television movie and in her autobiography (L. Gibbs, 1982a; see also Levine, 1982). As the national symbol of toxic victim turned dragon slayer, Gibbs is not in a class by herself. Rather, hundreds (if not thousands) of Americans have developed on the toxic training ground in much the same way as she, adopting a vigilant response rather than accepting the conditions presented to them.

Psychopathology and Exposure

The costs of coping with exposure must also be considered. The potential for long-term psychological impacts raises a number of questions. How does the psychological cost of toxic exposure compare with the costs of normal life stresses? Are there clinical symptoms of psychological harm? Are people negatively different after experiencing an exposure incident? These questions will be briefly examined in turn.

Comparing Levels of Stress. Fleming and Baum (1984) compared fifty-four residents from the immediate vicinity of the second most dangerous hazardous waste site in the country with residents from two similar New Castle County, Delaware, neighborhoods located at least five miles from any known hazardous waste site. Significantly higher stress levels were found among the exposed citizens than among control residents on a variety of measures.

Based upon these results, the authors conclude that residents exposed to toxic stressors, such as a hazardous waste site, may be at increased risk for stress-related outcomes. At the same time, they report that the stress symptoms identified are clearly subclinical, reflecting differences within the normal range but not indications of psychopathology.

Clinical Psychopathology. This conclusion can be contrasted with clinical research findings by M. Gibbs (1982). Reflecting on her study of Legler residents involved in the lawsuit, Gibbs (1982, pp. 37–39) stated:

> It seems to the author an inescapable conclusion that although many individuals in the litigation group may be well-adjusted, nevertheless the proportion of individuals with psychological problems, and especially the proportion of individuals with serious psychological problems, is much higher than one would expect in a comparable group which had not undergone the same stress. It is particularly impressive that these consequences remain two or three years after the period when most of the stresses occurred. The presence of serious pathology today, especially in areas other than health worry, attests to the power, pervasiveness and long-term nature of the stress experienced.

Further support for these findings can be drawn from Gibbs' later studies of toxic victims near two other New Jersey sites, one a landfill and the other the scene of a gasoline spill,[1] as well as from the work of Hertel on litigating neighbors of a Michigan dumpsite (Gruber, 1985).

Gibbs (1982) further posited that depression and other psychopathologies result from a loss of control caused by environmental stress. In fact, she found that the majority of Legler residents in her sample scored lower on measures of control than did comparison populations. However, a subset of the Legler residents were uncommonly likely to take control over events affecting them, supporting my finding that toxic victims who become activists have an overall net gain in control (Edelstein, 1982).

Personal Changes. There were additional negative changes evident in the Legler community (Edelstein, 1982). Some of these are briefly listed below.

1. Bad habits indicated behavioral manifestations of stress. An increase of smoking was reported for at least one member of nearly half of the sampled families. Other stress-related behaviors were identified, including nervous eating and insomnia.

One can question the extent to which smoking, eating, or insomnia were due to toxic exposure as opposed to other possible contributing causes. In fact, an interaction of contributing causes was likely. For example, one insomniac had fallen into a "mid-life crisis" about the time that the pollution became known. In thinking about new professional directions, he had begun to question his current job. The pollution incident further drained him of energy, making his job even less successful and satisfying. Not surprisingly, pressure at work began to mount, interacting with the intense pressure from the exposure incident and from his strained relationships at home. He reported being aggravated at work, an outcome that may have been additionally related to his inability to sleep (see Janis, 1971). In sum, his insomnia may have been due to any or all of these factors, or none of them.

2. Dreams about pollution also tormented some Legler residents.

I had bad dreams a lot. It was the same dream. The kids were in the pond. I'd always say, "Don't drink the water," but when they'd dive in they'd always drink some. In the nightmare, I'd see the kids floating in the water dead.

I shout out in my sleep all night long, especially after meetings. I fight with officials in my dreams. It started in 1978 and has gotten worse over time.

I'd have an occasional nightmare. I'd go into the shower and come out with open sores. There would be straight benzene coming out of the shower.

3. Changes in temperament were also reported, reflecting increased irritability and even a shift toward general hostility.

I'm antagonistic. Everything sets me off. I feel guilty about exposing my child. I fly off the handle more easily. I'm more intense than if this thing weren't hanging over my head.

Hostility was particularly heightened by the community meetings held to discuss the water crisis.

I'm only now getting my control back. Then I didn't have it. The only way to get it then was violence. My brain tells me that this isn't right, but I would have blown up city hall with the council in it.

4. Defensiveness in social relationships was a fourth factor. Some residents reported an inability to relax with others and a paranoid suspicion of their motives.

I'm not as nice a person as I used to be. I was trusting and patient. I was willing to listen to the other side. Now I listen, but I'm quick to be judgmental.

5. Depression was another outcome.

My daughter would get depressed and start drinking the polluted water. She'd say, "I don't care if I get cancer!"

I tried not to let it bug me. Some days I could ignore the problem; some days I got too depressed—I almost had a nervous breakdown.

In some cases, depression led to a loss of motivation and lethargy. It was as if by ignoring much of the world, one might cope.

Ever since the water problem, he's been unhappy with everything—his job, this place. He is melancholy and pessimistic. He thinks that things won't work. If something minor goes wrong, it's as if he says "Let's all lay down, we're going to die anyway." His energy comes in spurts. Last summer, there were two weeks in which he didn't want to do a damn thing.

I used to grab my wife when I heard a rhumba on the radio. Now I'm too tired. We don't go out like we used to. I don't dance anymore.

6. Guilt and self-blame were additional symptoms of stress. About half the respondents blamed themselves, not for causing the situation but for getting themselves into it and staying there.

We feel guilty that we just didn't leave, particularly when friends say we should have. Am I so terrible that I made my kids stay here under these circumstances? But to leave is to give up.

I felt guilty because I talked my husband into buying this place. We brought the kids here and poisoned them. Can you imagine how it must feel to stick your kid in poison? Incredible. What did I do to them?

Other stress symptoms also appeared. For example, somatic problems, such as migraine headaches and skin rashes, were widely reported. The convergent study by M. Gibbs (1982) provided clinical evidence for many of these changes identified in my qualitative interviews. She found comparatively high scores on indicators of health concern, an above-normal amount of hostility toward authority, and clinical levels of paranoia and depression.

Seeking Help

These findings suggest that individuals were highly stressed in Legler. Half of M. Gibbs' respondents indicated problems that normally dictate psychological counseling, with a quarter being in serious need of such help. While counselors might assist toxic victims to cope with their stress, both Gibbs and I question the applicability of conventional psychological counseling to a situation in which causes are so clearly environmental. Given the circumstances, true abnormality might involve the failure to show symptoms of stress (M. Gibbs, 1982).

None of the Legler residents reported seeking help from professional counselors. While this may reflect a general cultural bias against therapy in the community, there were specific reasons why psychological help was avoided. Legler residents feared that anything divulged during therapy might be exposed because of the lawsuit. Residents also distrusted the community mental health agency because it received funding from local governments.

Clergy also did not appear to play a major role in counseling residents. Religious residents drew strength from prayer, not from pastoral support.

I never told the clergy, but I prayed. I believe deeply that help through this period was from up above—He is the mightiest!

Personal religious belief provided a means for rationalizing toxic exposure and served as an important palliative coping tool. This was evident in Legler and every other instance that I have observed religious belief in contaminated communities. By focusing upon other-worldly concerns, the trials of this world became substantially less important. During her study of a minority community affected by pollution from a New Jersey landfill, M. Gibbs was told by residents that "they could not have dealt with this problem without God."[2]

Summary—The Individual

The stress of toxic exposure stems from the difficulty that the individual has in exercising control over the situation. Once touched by contamination, people have no definitive remedy available. There is no painless way out of the situation. Denial allows for a psychological balance to be maintained but does not address the external risks. Action involves recognizing the risk, a frightening prospect in its own right.

In communities such as Legler or Love Canal, most residents were part of a relational web including their families, friends, and co-workers. These relations influenced the extent of the victims' stress as well as their coping strategy.

Coping with Exposure: Couples

Given the relative isolation and mutual dependence of the couple in a privatized relationship, successful coping often depends upon the supportiveness of one's mate. For most relationships, coping with toxic exposure involves a mix of destabilizing and stabilizing characteristics. The Legler case study will be used to review these.[3]

Destabilizing Characteristics

The toxic incident in Legler was immensely trying for the predominantly young parents who lived there. In combination with other stresses confronting these couples, this added tension may well have contributed to the dissolution of some relationships and caused problems in others. Several residents independently estimated that as many as eleven divorces and twelve near-divorces occurred in the neighborhood during the period of crisis that followed the announcement of the pollution in Legler. It may be useful to examine some of the key conflicts that arose among couples during this period.

Blaming. Spouses sometimes held their mates responsible for getting them into the situation or for their coping strategy.

> *This didn't draw us together at all. We would have fights in which we'd blame each other for getting into Jackson. He had broken his promise that we would never have to move again. It was his fault.*

> *We would have awful fights. He would blame me for liking this house. He felt trapped. When he overreacts and wants to send our daughter to my sister's [to get her out of Legler], I miss her and he accuses me of overreacting.*

Such blaming was not always overt, and it sometimes smoldered below the surface for a long period of time. One Legler woman surprised her husband during the interview by admitting,

> *There were times I blamed him for the problems. I never told him about it. I'd think about it while I was hauling water around. I didn't want to live here, and he did. I'd particularly blame him if the cat got into the water, or I bumped into it.*

As this quote suggests, much of the resentment surfaced over lifestyle impacts, such as those entailed by having water delivered.

Many housewives also resented their husbands' ability to escape to work during the day and thus avoid water-related problems. Their separation also at times made the husbands unappreciative of these problems. For example, in one home where the well water stained clothing and dishes, the husband commented,

> *I didn't want to hear about the problems; I couldn't do anything about them. I was tired of hearing her complain about the clothes being ruined. We couldn't afford to buy new ones. It was frustrating. I looked like a slob. My mother said, "Doesn't your wife use bleach?" We'd argue about the dishes [which were stained from the water] or whether to buy a new dishwasher—they would just stain too!*

In part the blaming and anger stemmed from the inherent tension in the situation.

> *There was a lot of push and pull on us. We were so mad and aggravated that we'd yell. I'd yell when I was discouraged; we'd blame each other. The water was on my mind. The kid would be up all night itching from a rash. I just wanted to yell at someone.*

Distraction from Supportive Participation. Beyond the blaming, the stress of the incident was itself distracting. In some cases, one spouse lost interest in joint activities formerly important to the relationship.

> *He stopped wanting to go dancing; we never go to clubs any more. We fought like cats and dogs. I almost got a divorce.*

Not surprisingly, sex lives also suffered during this period.

Pressures to become active in the community organization also caused stress for couples. For a family to be represented and informed, some involvement was demanded from at least one of the mates. This new role invited various tensions, as when a husband represented his wife at meetings.

We usually argued after he returned from a meeting. He'd come home and tell me about it. I'd lash out. I expected him to know all the answers.

In other cases, one mate resented the others' inactivism.

I'd yell and try to get answers at a meeting, and he wouldn't say a word. Why didn't he do his share of screaming? I started screaming at other people too—like the bank about our mortgage. Why didn't he do his share? It might have been good if they'd heard a man's voice.

In all but one of the families of the board members of the Concerned Citizens group, marital strife resulted from the immersion of one spouse in a leadership role within the community group. The tension stemmed from the continual absence of the activist mate, the constant intrusions upon the home as it became a key link in the community's communications network, and the pressure and tension brought home by the leaders.[4]

We'd fight. He wanted me to quit Concerned Citizens because I was always bothered and upset.

I blamed him for being too involved in Concerned Citizens, for never taking a day off. He is chained to that committee.

Our home became a battleground. People would come in for help; they'd blame the executive committee for things. He'd be gone, and I'd be home all alone. He brought what should have been said to the township committee home to me. For a period, we just stopped discussing our problems.

Multiple Pressures. Also contributing to the destabilizing impacts of exposure were other existing pressures within the family. Multiple and simultaneous sources of stress might overwhelm a couple's resources for coping. For example, in one family where a divorce had occurred, the water crisis was felt to have "aggravated the situation."[5]

Stabilizing Characteristics

Not all Legler couples experienced relationship problems during this time. Several factors contributed to stability.

Lifecycle Factors. Within the interview sample, the most stable couples tended to be elderly, having the experience of long marriages to help them mitigate any problems. They lacked reasons to blame each other, nor were they as stressed by health concerns. Stable couples had generally lived in the community longer, avoiding the adjustments of moving, owning a first home, having young children, encountering problems with builders, and handling all of the other pressures that confronted the younger couples who were newly arrived.

For younger couples with children, the responsibility added to the stress. However, children often were cited as providing a reason to persevere.

> *Each time I had a problem with the kids and I couldn't cope anymore, I'd think I was going to have a nervous breakdown. But at night I'd feel that if I didn't have the kids it would be senseless to go on.*

> *We would fight and blame each other. Our son was a help because he was a joy. At least we had him.*

Coping Style. Beyond lifecycle stability, the likelihood of tension was influenced by the couple's coping style. A "laid-back" approach to dealing with stress was discernible among couples who avoided relationship problems.

> *We're calm, slow to get bothered. We have no choice. Why get upset over something that we have no way of doing anything about.*

Some couples who fought over the pollution incident were able to keep the conflict productive, for example, releasing tension verbally without hurting feelings. Other mates had complementary styles of handling stress, differing in degree of optimism or excitability, providing balance to the relationship.

> *Sometimes it was hard to accept, like when the container leaked and the floor came up. I punched the walls, I was so pissed. I felt stuck; it was beyond my control. My wife calmed me down and convinced me that it would all work itself out.*

Gained Confidence. Despite the tensions it produced, the water crisis also at times seemed to bring a closeness to couples that comes only from working through difficulties. For a family to make it through so complex and difficult an experience, there was a need for disclosure and sharing, for the development of trust and the willingness to learn new things about each other. Couples who succeeded seemed to have developed a resiliency and a stronger relationship.

> *Nothing can bother us now. In some ways this brought us together. We feel precious to each other.*

Long-Term Issues

Once the new water system was installed and there was a return to relative normalcy in Legler, stress from the pollution incident continued to surface for some couples. One reason involved finances.

> *When we had the water system put in, we had to go into arrears on our mortgage. We can't catch up. It's harassing us.*

Additionally, couples might even revive the tension.

> *For a while we had argued every day; then it was only after meetings. Now we only argue when someone reminds us about the water. Even with good water, the feelings are still there. We still occasionally have the same old argument, "If you would have listened to me!"*

A continuing source of concern involved fear of genetic and teratogenic effects on children. For caring parents, nothing is more unsettling than threats to their children's health. This lifescape impact also affected expectant parents. Kleese (1982, p. 3) provides insight into the psyche of the expectant mother acting as a mediator of her child's level of environmental hazard.

> One of the true enlightenments of pregnancy is the realization that the mother does not merely bear a child—as she might bear a load of books or wood. Instead, the mother represents the environment of the child; this situation is perhaps most analogous to an individual's relationship to the ecosphere. The mother is not merely a setting within which the child reposes, but she is a sphere of life wherein all of the interactions of natural systems necessary to sustain the child occur.

Accordingly, it is not surprising that these fears colored how Legler couples viewed the birth of their children.

> *We argued over having another kid. I refused unless we first had genetic counseling. If it weren't for the water, we would have conceived another child.*

> *We are afraid to have children because of the possibility of chromosome damage. The baby we just had wasn't planned. We worried during the pregnancy. The doctor did tests. When the doctor came out to announce the birth, my husband asked if it had all its fingers and toes.*

Whether the stress of a worrisome pregnancy affects the fetus is uncertain. What is clear is that the decision about whether to have children is one of the most agonizing legacies of toxic exposure (see also Levine, 1982).

Once born, the child is subject to continuing parental concern about the next generation.

> *I'm deeply worried about my sons genetically. They still have their lives ahead of them. They drank the water since 1971 and they swam in our pond. I told my sons to use genetic counseling. I don't want them to have the heartache of a deformed child.*

> *I'm concerned for my children's children. They have a cousin who is mentally retarded. This may happen to their kid someday; the children know this. I'm scared to someday be a grandfather—what will be born? Do you ask your kids not to have kids?*

Parental concern for children is the strongest factor motivating response to the announcement of contamination (see Chapter 3). Ironically, in communities such as Legler, parents commonly had moved to the suburbs in search of an ideal place in which to raise their families.

Coping with Exposure: Children

During a pollution incident, children are confronted by virtually the same lifestyle and lifescape impacts as are adults. While it is difficult to conclude that they are affected in the same way as adults, it is clear that they are stressed by the experience. Their stress comes largely from two sources. First, parental worry is passed along to them, as are the tensions due to parental stress. Second, children have a variety of experiences of their own involving direct impacts from waste sites, peer pressures, and the consequences of being taught to fear. The result is the sensitization of the child to the issues involved in toxic exposure.

Parental Sources of Child Stress

As members of a family dominated by the concerns of adults, children experience much of the stress of exposure incidents vicariously from their parents. They must adjust to their parents' attempts to cope with the situation. There may be no more important indicator of the child's emotional response than the success of the parents' efforts to cope (Freedman, 1981).

The Effects of Parental Stress. In locales such as Legler, the high level of community stress resulted in a corresponding level of intrafamily tension. Beyond the strain on marriages, there were instances when children took the brunt of the parents' frustration.

Before, I was very calm and in control. But during this time, it didn't take much to set me off. My nerves were shattered. If the kids dropped a drink, I'd fly off the handle.

When I was depressed, I'd take it out on her. I'd pick on her. We'd have screaming contests. At least I can say that I'm sorry.

Children of community leaders paid the costs of activism much as did their parents. Their homes bustled with activity. The entire family was under stress. As one Legler leader recalled of her children,

There was nothing normal. The phone rang, there were meetings we had to attend, there were reporters. They [the children] were scared. They saw the tension and the emotion. It was far from peaceful and tranquil.

Activist parents might additionally interfere with the normal attentions that a child expects. Lois Gibbs (1982a, p. 53) offers an example of how the demands of community leadership interfered with her role as mother. Returning late from Washington, D.C., the night before her son's birthday, she not only forgot the birthday, but she ate much of the cake made for her son by the child's aunt. The next morning, her son was in tears. Quickly buying presents and a new cake around her busy schedule, she fit in an impromptu party around meetings with lawyers, phone calls, and preparations for an important meeting.

By now it was 7 p.m. and we had to be at the meeting by 7:30. So we lit the candles on the cake and quickly sang happy birthday. It may have been the only birthday party for a six-year-old that had four attorneys and two doctors as guests. Michael opened his presents and cut the cake; then the adults left for the meeting. I hope Michael will understand and that he will forgive me someday.

A community leader from another contaminated community told me that her young child was so accustomed to accompanying her as she engaged in her activist work that he would ask her in the morning, "What's our schedule today?"

The Effects of Parental Worry. Parental fears for children's safety put a damper on the joy of family life. In one case in Legler, a mother feared retaliation against her child because of the father's activism.

We worry that she might be harmed when she leaves the house. They might try to get even. Our fear has given her fear. Now she worries about the "bad man!" She is explosive like her father. She has trouble sleeping. She's not as

secure as she used to be. She doesn't like being alone or being away from her parents.

At times, total panic swept Legler. When one of the children died, some children were sent to stay with relatives. Moreover, Legler children were exposed to explicit communications that all was not well. Some couples argued and agonized about leaving their homes. Parental arguments sometimes contained explicit and understandable messages of fear.

Parental self-blame appeared to be a contributing factor in parent-child relationships in Legler. This was particularly evident in the behavior of a young mother during a lengthy interview in which she continually reinforced her son for clinging to her. The mother admitted,

I became possessive because of my guilt over having exposed him to this. I'm overprotective. I don't let him experience what three-year-olds usually do. I like to keep him near me.

The Child's Experience

Children also have direct experiences from an exposure incident. They are aware of many of the direct impacts of a facility, and they suffer some of the secondary impacts. They hear conversations and watch television. They interact with peers, teachers, and others with reactions to the pollution. And they confront explicit communications from their parents about their safety. Thus, even younger children, who lack the complex reasoning necessary to understand the full dimensions of toxic exposure, are quite aware of the disruption to a normal and secure home life.

Direct Impacts. In Legler, children's activities (e.g., playing outside, having company, and camping in the backyard) were limited by landfill traffic, by putrid odors, and by parental fear of vectors. Similar conditions can be found for children living near other hazardous sites. For example, a girl growing up near Al Turi Landfill, Inc. in New York State recalled doing her homework after school to the rhythm of the equipment noise. Her brother suffered landfill-related nightmares; the sounds of trucks arriving early in the morning would cause him to jump out of bed still asleep. He would awaken standing bolt upright. Children became increasingly sensitive to changes in the environment, observing the pollution of their favorite ponds and streams by leachate and litter. On their way to the school bus, the children had to walk a gauntlet of giant 18-wheeled trucks, taking the comments of drivers as a threat to their safety.

In such situations, the cumulative sense of intrusion exceeds the incremental effects of each nuisance. Normal life comes to be dominated by the facility. For children, whose entire world revolves around home even more narrowly than it is likely to do for their parents, the psychological threat is all the more powerful.

The transformation of a benign environment into a feared one represents a disconcerting constriction of freedom for a child. Thus, two young adults from Legler whom I interviewed recalled the rural environment that preexisted the mining site and the landfill some fifteen years earlier. It was a place where children made their own adventures, where time was spent exploring the local woods, following old stagecoach paths. Changes in Legler, in the words of one of the teens, made "my whole childhood disappear." Both remembered having to pick up trash thrown by people coming to use the landfill. One remembered that when he had stopped drinking the water shortly after the landfill opened, his father ridiculed his fears. It was not until the announcement of pollution, some eight years later, that the father appreciated his concern.

Teaching Children to Fear. As part of their adjustment to exposure, parents have to change their children's behavior. In Legler, parents literally had to teach children to fear the tap water. For older children, this message was not overly difficult to get across. But younger children who were able to use the faucet but not able to understand complex reasons for refraining from its use posed more of a problem. Parents were forced to use blunt messages that these children could understand.

> *We told them that the water was poison, like cleaners. None of the kids went near the water.*

> *We told the kids that there was poison in the water, that they couldn't even brush their teeth with it. One time our son forgot and drank some. He came out screaming, "I drank the bad water, I'm going to die now."*

Children, in turn, acted as agents to inform others of the problem, as this Legler parent explained:

> *They'd often tell people that our water is bad. If we had company, they'd say, "You can't drink that, it's poison." Then a guest drank some water one time; they expected to see him drop dead.*

Once they learned that tap water was poison, children next had to define the generality of this phenomenon.

> *If visiting my relatives, they'd ask if the water is poison.*

Younger children were additionally confronted by a confusing set of new rules that were not always consistently supported in Legler neighborhoods.

> *I can hear the kids talking outside. Some of the kids were allowed to use an aerated sprinkler. We let our daughter use it, but other kids were not allowed. This caused confusion. The kids would talk about how each parent says different things. The neighbors' children were so afraid to touch the water. Our daughter wasn't—she bathed in it.*

> *For a while, nobody played with our daughter. They would be allowed to go under the sprinkler and our kids weren't. They would taunt our kids with "ha, ha, we're having fun."*

When the central water system was later installed, Legler children were faced with learning that they could trust this source of water. As parents recalled,

> *Later it was hard to get them to use the faucet again.*

> *For us the faucet is like a new luxury, but for the kids it's like a new invention.*

There were other opportunities for children to learn about the situation. During the Legler interviews, parents often spoke freely of their concerns with their children present. This was not the first time presumably that the children had heard these concerns expressed. These were children who knew that all was not well, regardless of their ability to fully comprehend just what was wrong. The same thing occurred at other locales. A Love Canal boy, about 10, stood mutely by, shuffling his feet and averting his eyes from my video camera, while his mother described his various physical and psychological problems.

> *My son has respiratory problems; he has asthma; he has been sick all of his life here. He has been hyperactive. He has been on sedatives from the time he was seven months old to eighteen months old. And now he's having psychological problems associated with this because he had to change schools. My son refuses to go. We changed schools again, but then he had another bad attack of asthma.*

Peer Stigma. Children also learned of events through their peers, who were not necessarily supportive. In fact, peers engaged in the same patterns of victimization that confronted adult victims. One Legler teenager recalled,

At school, one kid told me he saw me on TV and that I was really stupid for moving into this area. Another said that he felt sorry for me. I hate both of them. I don't want pity. I was so mad I hit him.

Peer group dynamics played out in mirror image the conflicts that divided Jackson Township's adults. Thus, one Legler leader reported of her daughter,

She took abuse from other kids because of my involvement. She got punched on the bus; they called her names and told her that her mother is a troublemaker. "My dad said that your mom is going to make our taxes go up."

Similar dynamics are described at other sites as well, suggesting that children face a degree of social isolation as the result of toxic exposure incidents.

Media Coverage. Children were influenced by media coverage. It was particularly frightening for children of activists to see what their parents said publicly. Lois Gibbs (1982b, p. 11) describes her son's experience: "For example, picture yourself five years old watching the 6 o'clock news report. Your mother is being interviewed saying, 'There is dioxin in there. Dioxin kills. I don't want my babies to die.' What is the child supposed to think?"

Sensitized Children

Some children became highly sensitized to the possible dangers of water. During the Legler study (when a new water source had already been provided), I observed a young girl refuse to join her playmates under a sprinkler despite her parents' permission. Her mother explained that her daughter could not get past the belief that the sprinkler was dangerous. A few doors away, a teenager's parents reported that she continued to face away from the shower even though her home was hooked up to the new water supply. She was responding to the irritating characteristics of the original water source.

Legler parents told of other examples of children's sensitivity:

What hurts us is to see our six-year-old be paranoid—"Can I drink the water?" "Can I swim here?" "Can I bathe?"

When we go camping, the kids don't want to swim in a pond if "it looks polluted." If we see a 55-gallon drum, they say "look at the toxics." If there is a rainbow in a puddle, they say "looks like chemicals."

Toxic concerns may become a factor in a child's ability to sleep. For example, Freedman (1981, p. 624) describes very young children at Love

Canal who had "chronic nightmares of toxins oozing from their bodies, while others have regressed to bedwetting, baby-talk, and clinging to their mothers." Such worry might carry over into behavior at school, as this Legler parent recalled:

> *Our son was bothered by the situation. He wondered if someone made his little sister die. He was moody in the second grade. The teacher talked to him and found out that he was afraid of dying. He wouldn't study water pollution in class; he avoided the topic. The teacher said that he can pick out Legler kids.*

Both Freedman (1981) and Lois Gibbs (1982b) note that younger children at Love Canal feared premature death. Older children were more able to appreciate the abstract horrors that so affected their parents. They too adopted a view of future as promising evil. Gibbs (1982b, pp. 10–11) observed the effects that this had upon Love Canal teenagers.

> The young people already had normal teenage problems and strains, and the Canal made growing up just that much harder. Young women suddenly had to face the fact that they might have chromosome damage or that their eggs might already be damaged by the chemicals in their environment. They feared that they might not be able to carry babies or give birth to normal children. These facts were devastating to the young women with their whole lives in front of them, and especially in our community where most people marry within a short time after graduating from school and immediately begin their families.

In a similar vein, Freedman (1981, p. 624) writes of one of three child suicides at Love Canal in the summer of 1980.

> The 14-year-old who killed herself with sleeping pills, reportedly feared that she would develop cancer, as members of eight out of twelve families living on her block already had. She feared that it might be breast cancer, or a cancer that would leave her unable to bear children. She worried that she would be unable to have a normal marriage. So, at 14, barely into puberty, she took her life.

Although I know of no similar outcomes in Legler, teens clearly feared what toxic exposure might mean for their futures.

> *What will happen to me twenty to thirty years from now? I get scared about it.*

Conclusion to Children and Coping

In summary, children underwent many of the same lifestyle and lifescape shifts that were faced by adults. In some cases, they may have been buffered by a child's view of reality. In other cases, they may have had enlarged fears. In either case, they shared with their parents the disruption of toxic exposure.

Case Studies of Family Dynamics

Family stress due to toxic exposure continues throughout the time that the family remains exposed, with certain stressors extending even longer. At the onset of the Love Canal crisis, two-thirds of Levine's respondents reported a high level of family strain. The next year, when nearest residents had moved to new homes, nearly half of these families showed improved relationships (Stone and Levine, 1985). Therefore, removal from the situation can be seen as a key element in stress reduction. In the vast majority of toxic exposure cases, however, this removal does not occur.

As victims grapple with coping in their contaminated homes, family dynamics vary. Although contamination intensifies problems in some relationships, other couples find that coping with contamination draws them together. It would appear that a consensus within the family about the exposure is required if the family is to stay unified through the experience. The contrast between united and divided families can perhaps best be seen by presenting two case studies of toxic victims from different sites in the eastern U.S. I studied both cases in 1983 at the request of the lawyer representing the families in separate lawsuits. As with other toxic victims discussed in this volume, the names of the families have been changed.

Case One: Family Disintegration from Exposure

Louis, a retired man in his sixties with one college-age child, had been blind since his late teens. Some twenty-five years prior to our interview, he and his wife purchased a rural home with an adjacent rental property. The financial independence this property afforded allowed them to spend much of their time at home. Despite his blindness, Louis was independent to an extent amazing to a new acquaintance. He moved freely in the out-of-doors. His ability to describe the natural features of the area surpassed that of many sighted people. He actively participated in maintaining his property, doing such chores as splitting wood for the stove. Particularly since his retirement, his life had been centered almost entirely around the home and property.

After a chemical factory took over an adjacent farm-related industry in the early 1970s, Louis and his wife suffered various direct impacts from its operation. As the factory became an increasing preoccupation for them, the couple came to spend the bulk of their free time seeking help from government. Even when it meant neglecting their child, Louis would spend hours dictating letters to various officials who might help them. His wife typed the letters and mailed them. When groundwater contamination was later identified in their well, they felt certain that the factory was the source. As a result, they increased their efforts to get assistance.

Despite little success in eliciting help, Louis became increasingly preoccupied with the pollution. He believed that he and his family were exposed to dangerous chemicals that affected their health and might cause genetic damage as well. He saw himself as fighting a wrong brought not only against his family but against the entire community. For Louis, the key issue was his family's exposure to chemicals and the inevitable consequences.

In contrast, his wife began to rethink their obsession with the factory. She felt guilty that her child had been raised during protracted turmoil that prevented a normal home life and caused the boy mental strain that had become evident in his teenage years. But to Louis, his son's mental strain resulted

> from the physical problems from the water. His nerves are affected from ingesting the water by drinking, eating and bathing.

The wife had also started to develop other interests. She was "reborn" through religion and also undertook a career. As she and their son became increasingly mobile and centered elsewhere, the pollution issue became correspondingly less important for them, even as it became increasingly central to Louis. Continual misunderstanding resulted from this different emphasis. Thus, at the time they were interviewed, in 1983, Louis' wife and son maintained that the incident was over and that it was time to get past it and go on with life. But to Louis, "you can't walk away from something that is now within your body." Rejecting their attitude as a "Band-Aid," Louis became increasingly bitter at his wife's lack of interest in pursuing a resolution of the case. For himself, he proclaimed,

> I don't give up. If you're going to beat me, you're going to have to kill me.

To fight for restitution of what was lost had been cathartic for him. As a blind man, he was particularly dependent upon his wife for the ability to wage this effort. He had been able to keep up his high level of activity because she acted as his secretary, collaborator, and chauffeur. But as his wife refocused her energies, Louis was increasingly hampered in his ability to fight. Sitting alone much of the day, he was unable to distract himself through reading or traveling. Instead, he was left to stew in anger.

Increasingly, he persevered on actions that he felt were necessary, not relaxing until they were completed. For example, at the time that I interviewed his family, a form sent by their lawyer had sat uncompleted for a week. Louis was obsessed with getting the form finished. He was outraged that his wife had not helped him fill it out immediately. He reminded her of it continually, further alienating her. It became obvious that as the pressures had mounted in the family, Louis had not only lost his secretarial support, he had also lost his emotional support.

> *It's hell for me. I'm here all the time. I cannot leave. She refuses to read and write for me. She's out all day. When she comes home, I'm in bed. I eat sandwiches. I'm like a prisoner. I have no nourishing meals. It's not a normal relationship. She eats standing up in a rush all the time.*

As his wife avoided the issue, she needed also to avoid him. He was left thinking about what needed to be done and composing letters that he had no personal power to send. From his perspective, Louis was caught in a true dilemma. His life was guided by his desire to respond effectively in an ongoing battle to protect his family and community. To do this, he needed his wife's help. When a response was required, he reminded her. Seeing her but infrequently, the topic dominated his communications to her, motivating her to avoid communicating in return. This dynamic is illustrated in the following exchange from their interview.

> Wife: *Louis always mentions pollution. I went along with that that for a long, long time.*

> Louis: *My only regret is that I don't have more time to devote to this. It's a necessity.*

> Wife: *He talks about it all the time. Every time he sees me he says, "We have the letter that has to be sent."*

> Louis: *The percentage of time devoted to this is about two minutes a day on the average.*

> Wife: *Do you deny that every time we get company, two-thirds of the conversation is about the factory?*

> Louis: *That's an exaggeration—it's a lie!*

Wife: *I'm tired. We've been sending letters since 1974.*

Louis: *I'm stronger than you are. I'm a human being too. I regret that I don't have more time to spend on it. It's wrong to let him [the factory owner] get away with it. I only devote two minutes a day, on the average.*

Louis was being prevented from dealing with the issue that consumed him, but to mention it was to make it even less likely that the issue would be dealt with. From his wife's perspective, creating a normal home life required that the topic of pollution be expunged from their interaction. Yet the more she fought to eradicate it from the limited family communication, the more obsessed Louis became. He even began to lose the powers of concentration that had allowed him to organize his thoughts despite his blindness. Once proud of his competence, he became increasingly embarrassed by its loss. He also begun to suffer from periodic choking episodes that came on during periods of extreme frustration.

Although Louis sought justice, revenge, and relocation, which would further provide him needed confirmation of his perceptions, his wife and son just wanted the crisis to be over. They accomplished this cognitively—it was over because they decided it was over. Their view was that even if the pollution issue never got resolved, it was more important to bring something positive into their lives. This fragment of conversation captures their conflict:

Louis: *It's a scientifically proven fact that there is chromosome damage.*

Son: *I realize it. But there is nothing there now!*

Wife: *You can't only worry about physical effects, but mental effects also. We could go crazy fighting all the time.*

Son: *Physically I was young, so it didn't do much to me.*

Wife: *You can't expect a child to go through this.*

Louis: *It's not a manufactured thing. You have to try to do something.*

Son: *The situation is gone.*

Louis: *You can't put a paper bag over your face and say it is not there.*

Wife: *You put holes in the paper bag and keep on going.*

Louis: *That's silly.*

Son: *That's not silly.*

Their coping strategies were totally divergent, although neither Louis' fixation nor his wife and son's defensive avoidance was particularly adaptive. The wife and son sought to make up for the years lost due to the incident; Louis couldn't get past it. The wife, feeling guilty for subjecting her son to so distorted a home life, was just as fixated on eradicating the issue from their lives, no matter its merit, as Louis was on keeping it as the central issue. This difference became a barrier between them. Louis claimed that he wanted his wife to leave; he appeared to have lost his trust for her. She wanted to stay until their son was older. Both blamed the pollution incident for the destruction of their relationship.

Were there causes for the schism in this family besides the contamination? Louis' wife might have evolved toward interests separate from her husband's in any case, leading to her increased independence.

In this case, one can conclude at the very least that the pollution issue contributed strongly to family conflict. Louis' wife appears to have been driven to other interests as much by her avoidance of the pollution situation as by her attraction to new interests and independence. It is not surprising either that she evaded her husband or that Louis interpreted her evasion as a break in trust. The wife's changed role was more than a liberation from the homemaker role or rebellion against the dependency of another; it had the added significance of ending the years of shared collaborative activity against the polluter that had become their dominant context for relating.

This case illustrates how a family can be split by the stress of a pollution incident. For Louis to persist in seeking answers to the fundamental questions of health and justice was quite reasonable. And yet, because our society is so ill-prepared to tackle these questions, he was stalemated in pursuing them. His task was Sisyphian. What appeared at first to be a competent and direct approach to the family's needs thus became an unproductive and dysfunctional element of the family's life. In contrast, his wife and son chose to avoid this unrewarding focus and to reorient their lives to more constructive purposes, a healthy decision from the standpoint of their mental health although it avoids dealing with persistent issues of their experience. In a sense, therefore, neither choice is totally healthy. In fact, people in this family's situation may not have a totally healthy choice open to them.

Case Two: Mutual Support
in the Face of Toxic Disaster

Sonia and her husband had enjoyed their rural life together despite the constant infringement of the neighboring waste site. But when their well was discovered to be contaminated by leachate from the waste site, their ability to adapt to an adverse situation was strained.

Despite this adversity and the impact that it had on their attempt to live a simple life close to the soil, they sought not to bother others or to be excessively disturbed. Their coping strategy drew heavily from their religious faith, as well as from a conscious effort to appreciate each day and to maximize the positive. Sonia commented,

> *You know, we're studying the Bible. And there were certain phrases we picked up. If you think good thoughts, they will multiply. If you think bad thoughts, they will multiply. You talk about these bad things, and it spoils your whole day. I won't be able to fall asleep tonight because all these bad thoughts are on my mind. So we try to dwell on good thoughts constantly.*

Reflecting this approach, they refused to hate their neighbor, the polluter, despite his responsibility for their predicament. Sonia noted,

> *All my life, the kind of policy I work with—if a person has one percent of goodness in him, appreciate that one percent and forget about the rest. And this is the way I lived all my life. And I believe Erik is the same way. I don't hate [the polluter]. He's doing this for a living. If this is what he went into, it's his business—I can't be critical of that.*

And Sonia's husband, sharing this ability to forgive, said of the landfill operator,

> *He's a genius. He only should be pushed in the proper direction. Like a bank robber who can cut metal—he would make a great EPA guy.*

Both share a profound sadness at the change that has overtaken them because of the pollution. Sonia is severely depressed; her husband is equally despondent. Yet their shared understanding allows them to be mutually supportive. They actively help each other cope with the situation. In contrast to the prior family, coping with toxic exposure has brought them even closer together.

Stigmatized Relationships—
Outsiders Just Don't Understand

Outside the immediate family, toxic victims frequently find that exposure changes their relationships with friends, relatives, co-workers, and others not in their circumstances.

Friends and Relatives

Beyond the nuclear family, our relational web encompasses a range of friendships and associations with kin. Under normal circumstances, and in the face of most major life crises, this relational web can usually be counted on to provide the individual and family with various kinds of help, aiding in the attempt to cope with the crisis. Thus, while such networks may serve as a source of social stimulation, they additionally provide for socioemotional support (see Unger and Wandersman, 1985). But how does this support network function in the context of toxic disaster? The Legler case study leads to various conclusions (see also Levine, 1982).

As a sprouting development within the New York metropolitan area, Legler was readily within reach of the major urban centers from which most Legler residents had moved. Many maintained close ties with friends and family, exchanging frequent visits on weekends. In most areas of Legler, where intraneighborhood privacy norms were strong, the prior network of relationships continued to be primary while neighbors were kept at an "acquaintance level" of association.

Friends and relatives assisted victims in a variety of ways. They brought them water; some offered financial assistance; several said that they would take the children until the crisis was over; some offered the use of showers. Yet despite such actions, most residents felt that, relative to their needs, they had received very little help overall.

What might account for this perception? To begin with, it was as unclear to residents as it was to their friends what concrete actions might be helpful. Sympathy was usually offered and at times it was very helpful, as when parents reassured their adult children that they should not blame themselves for buying a home in Legler. But often mixed with the sympathy was the contradictory message, "How can you keep living there?" Residents already had conflicting feelings about choosing to stay in Legler; such "support" made them either defensive or doubtful. Sometimes, blatant criticism was offered, arousing parents' guilt for staying in Legler or at least not removing their children.

Rather than being intensified as a support network, friendships often weakened because of the situation (see also Stone and Levine, 1985). In many cases, people who had frequently visited Legler before the water crisis just stopped coming. These recollections are illustrative:

> *Only my mother will come here. It's like we have the plague the way they avoid Legler. They feel that they don't want to expose themselves to our house. My wife's parents don't even want to know about it.*

> *Before the water, we entertained every weekend. Now only immediate family come. The others are afraid. We're afraid to invite them. People even called*

*to ask us whether they were exposed to pollution from our water during their
visits. This drains; it doesn't help!*

As these quotes indicate, friends and relatives at times showed self-
concern as much as support.

Toxic victims sometimes helped close relatives pretend the problem
didn't exist. Thus, during one interview, the wife's parents sat watching
television upstairs while the couple described their experiences. The
couple didn't want her parents to know how bad it had been.

Sometimes when parents knew of the situation, they were not perceived
as being supportive. As this young Legler father spoke of the pressure
he received from his family, he clenched his fists and assumed an angry
look and tone.

> *People act like we're diseased. They say, "If I was you, I'd get out of there."
> But where are you going to go? Everybody is a doctor or a lawyer! They're
> not in our situation. I'd get upset with them and say that the problem is
> being looked into. People say, "If I were you, I'd sell my house." The water
> is fixed; we went through the tough part already. My relations are not
> supportive; they always criticize—"sell the house; get out." They don't know
> the situation! We couldn't afford an apartment and a mortgage. But they won't
> accept this explanation. Our water stained all our clothes. My mother would
> ask, "Do the kids have any unstained clothes?" Then she'd buy the children
> "nice outfits."*

For one couple who moved to Legler for a back-to-nature existence,
parents reinforced their prejudices about how their children lived using
the pollution incident.

> *Our parents hate our lifestyle. They think that this place is a step down from
> our former bi-level. The water just emphasizes our mistake in moving. It
> reinforces their belief that this was a bad idea; they won't accept our lifestyle.*

Furthermore, as residents were interviewed by the media, they became
objectified by friends and relatives as "celebrities."

> *With the newspaper picture, no one ever wanted to let me alone. Friends
> would ask about it; they'd make me lose my concentration at bowling. They'd
> get nervous when I'd talk about it. They didn't really understand what I went
> through.*

This notoriety also invited the suspicions of friends (see Stone and
Levine, 1985).

The theme of "outsiders don't know what we went through" is echoed in virtually every toxic exposure incident I have investigated. It is as though the victims share a private experience that can only be known to those who have undergone it themselves.

Distancing from friends occurred in another way as well. Thus, friends sometimes got tired of residents' preoccupation with their problems.

> *Our friends say that they don't want to be with us anymore because we are so caught up in the water problem.*

Ironically, during the Legler study, the only family not reporting a change in relationships was an elderly couple undetered by the pollution from continuing to use their well water. They reported,

> *Everyone who comes here even now wants to drink the water. We have friends here all the time. They're not concerned about the water.*

This suggests a mirroring effect. Residents who believed their water to be dangerous may have convinced their social network of this. The reverse appears also to be true. Ironically, for those disbelieving the severity of the incident, social support among friends and relatives may have been greater than it was among those for whom this was a true and complete disaster (see also Fowlkes and Miller, 1982).

Co-Workers

For the men and few working women of Legler, work provided a major life context outside the community. At times, this helped victims' ability to cope by giving them a space separated from the crisis and an opportunity to have an outside perspective on the situation at home. But this separation was often undermined by co-workers and others during the course of the workday. Although generally intended to be supportive, inquiries and comments tended to break down the separation and deprive residents of their escape from the pressures of the crisis.

> *My co-workers and people I deliver to would ask about it; it was nice, but it dwelled on my mind that way.*

In some cases, co-workers ridiculed the victim.

> *The other workers would laugh at me for overreacting. You're standing there and being hurt by it.*

> *A few made jokes like "If you need a light, take a bucket of _____'s water; it glows."*

As it did with friends and relatives, the inherently private experience of the problem made it genuinely hard for outsiders to appreciate the severity of the situation (see also Levine, 1982).

> *Most people were sorry that it happened, but they didn't appreciate the problem.*

Other Outsiders—Stigma

Barton (1969), in discussing all disasters, notes that blaming of victims tends to occur if the impacts of the disaster are specific to one group, if the impact occurs gradually, if others have vested interests in the causes of the victims' deprivation, if media content suggests blame, if others have little personal contact with the victims, if the blaming dynamic becomes widespread, and if altruistic values are absent. These conditions characterize the toxic cases discussed.

In Legler, stigma effects were felt when Legler residents encountered others from outside Jackson Township. Central to stigma is the observability to others of a discrediting feature, what Goffman (1963) terms the "mark." Legler residents felt that their very identification with the contaminated community marked them for stigma.

> *If someone sees I'm from Jackson, they ask, "Are you from that water area?" It's terrible.*

Neighbors of landfills at several sites reported that others identified their homes with proximity to the landfill.

Stigma and blaming the victim were key elements of the community conflict surrounding the Legler pollution incident, as suggested by these comments:

> *Every place I go on the other side of town, they mock us: "There's nothing wrong with your water; you're only out for the money."*

> *The local papers say you can't let two square miles affect the township. This hurts. We didn't create this situation.*

> *They don't give two hoots. People at the town meetings tell us to shut up, that they're sick and tired of hearing us and that we take too much time up at meetings.*

> *I saw people on the other side of town who said that their well isn't polluted so why should they have to pay for mine. They blame us for all the news coverage convincing people not to move here.*

As reported earlier, township officials focused upon overcoming the stigma directed at the entire town because of the Legler contamination. But as the town acted on this goal, residents of the Legler section were further blamed and stigmatized.

Similar dynamics have occurred elsewhere. At Love Canal, for example, Fowlkes and Miller (1982, p. 98) describe stigma directed at those believing in the contamination by those who did not. The latter group viewed themselves as having "normal identities," while the believers were seen as abnormal and thus illegitimate. In this way, the believers' perception that the canal was hazardous, so threatening to the disbelievers, was also characterized as illegitimate.

In the highly divided community of Centralia, Kroll-Smith and Couch reported[6] that Centralians were stigmatized. Reflecting the view that they were not able to handle their own problems, others made them the butt of jokes. For example, overheard in a bank was the comment, "Here comes a Centralian, the money is hot!"

Neighbors—Proximate Support

As noted previously, Legler was a suburb where neighbors knew each other; they were friendly but did not necessarily become close. Exceptions existed on both extremes, with relationships in one cul-de-sac torn by frustrations with the developer while another section shared almost a communal life. But it was not until the isolation of Legler in the throes of the water pollution crisis that neighbors began to develop close relationships. These relationships were particularly helpful for many residents in coping with the disaster because of the distancing of friends and relatives formerly central to residents' social networks.

Thus, if the central problem in gaining continuing support from friends and relatives outside of Legler was the seeming inability of those not sharing the situation to understand it, then neighbors who shared the same situation might now offer each other significant social support. Much of this support came simply through conversation. Because they were in the same predicament, neighbors frequently were willing listeners for each other. They could test their ideas, discuss their fears, and seek empathy. This sharing gave them the opportunity for both emotional release and supportive feedback. They were also able to gain a basis of comparison with others whose situation was similar to theirs.

> *Neighbors were very helpful. They were supportive in every way. I'd go to talk to them over a cup of coffee; then I'd feel better.*
>
> *I could let my aggravation out. I wasn't alone. And when I'd see others' problems, mine didn't seem so bad.*

Neighbors also traveled together to the meetings of the Concerned Citizens Committee. This afforded them the opportunity to discuss options and impressions. They were not involved in a complex and frightening situation alone.

Despite the importance of neighboring to the overall climate of social support, there was not a total consensus. Some residents did not believe that the pollution was important. Others did not approve of the community organization bringing suit against the township. The divergent views contributed to ongoing tension in parts of Legler. And even where social support was initially effective, over time the internal obsession with the crisis became counterproductive for some residents. Some noted that all that they ever talked about with their neighbors was the water, "even when we're tired of talking about it." Sometimes this intensified anxiety, as occurred with one woman who became so nervous that her husband forbade her speaking with neighbors.

Considerable variation existed within sections of Legler as to the development of close neighborhood ties. Thus, in some areas help came readily; in others it had to be requested. Note the differences between these comments:

> *You could borrow water from your neighbors, but you had to ask, they wouldn't just offer.*

> *People borrowed water if they forgot to put the flag out. We exchanged spring water. We were lucky to have good neighbors. This helped bring us close together. We would not be as close now if it were not for our water problem. We wouldn't know many people in the community. We're like a family.*

Overall, the interdependence among neighbors added a dimension of community spirit to Legler (see Chapter 6). This togetherness was demonstrated when the new water lines were ready. In several sections of Legler, neighbors worked together to dig the holes for hooking up the city water. Strong feelings of cohesiveness, such as these, were expressed:

> *All of the people were in the same boat. I love our neighbors. I fight more for them than for us.*

> *We felt like an "us."*

This neighborhood dynamic helps to explain Stone and Levine's comment based upon their Love Canal analysis that only one-third of their residents reported losing friends while half reported making new friends (1985).

Summary—Individual and Family Impacts

Toxic exposure places the adults within the family under particularly severe stress. Successful coping depends in large part on the ability of the couple to avoid being destabilized by these pressures. However, the family is likely to be isolated from friends, relatives, and co-workers from outside the affected area, who do not share the perspective of the victim. Those outside the victims' relational web are even less likely to be supportive. Particularly if their vested interests are countered by the pollution event or the demands and actions of the victims in coping with it, these "strangers" may form a hostile group of outsiders, frequently arguing that activist victims are using the situation for their own ends (see Stone and Levine, 1985; and Fowlkes and Miller, 1982). Meanwhile, the institutional network is unlikely to provide relief, disappointing expectations that government will step in to solve the problem. In the face of the resulting isolation, the support given by neighbors becomes increasingly important. Such proximate relationships are the basis for the development of a sense of community, as people combine forces in the face of disabling relationships with government.

Notes

1. Private communication from Margaret Gibbs; see also Gibbs, 1986.

2. Private communication from Margaret Gibbs; see also Gibbs, 1986.

3. I have not had much direct experience with single adults or single parents in contaminated communities. My Legler sample consisted almost entirely of couples, as have my samples elsewhere. The question of coping with contamination outside of a relationship deserves further attention.

4. It should be noted that such stress due to involvement is true not only for toxic activists. Thus, Holman (1981, p. 147) suggests that "up to a point, community involvement helps a person improve his or her performance of marital roles; however, as the involvement begins to take increasing amounts of time, marital role performance declines and there is a concurrent drop in marital satisfaction."

5. Stone and Levine (1985) note of their Love Canal analysis that coping resources, such as higher income and educational levels, helped to buffer some residents from such stresses.

6. Conversation with Steven Kroll-Smith and Steven Couch, August, 1984.

5

Disabling Citizens:
The Governmental Response
to Toxic Exposure

With the discovery and announcement of contamination, toxic victims suddenly find themselves in a complex institutional context made up of the various local, state, and federal agencies having jurisdiction over their contamination incident. This is an unfamiliar life context for most people, one for which they lack experience. Their lives are, in a sense, captured by agencies upon which they become dependent for clarification and assistance. The essence of this relationship is depicted by Levine (1982) in her description of how Love Canal residents perceived their treatment by the New York State Department of Health. "Because DOH officials did not pay serious attention to the task of providing information to them and working through the implications of the information, the residents felt that they were being treated not as rational, respected adults but rather as though they had somehow lost their mature good sense when they became victims of a disaster they had no way of preventing" (p. 74).

Effectively, toxic victims become "disabled," as suddenly they are dependent upon professionals to expertly handle various areas of life formerly governed by their own naive wisdom (Illich, 1977). What is lost is their ability to participate directly in understanding and determining courses of action important to their lives.

For example, the regulation of environmental risks has come to embody the pseudotechnocratic character of modern society. The criteria by which decisions are made do not reflect social values expressed through the political process, but rather political decisions hidden behind the rationale of technical standards made by experts (see Habermas, 1970). Thus, the question of acceptable risk has little to do with people's values, but much to do with the economic and political forces concerned with the costs of environmental standards. Government professionals control resources and make key decisions regarding the response to toxic incidents. In the shadow of these political and economic forces, there is additionally a much-touted gulf between the "subjective" concerns of impacted citizens and the "objective" assessments of professionals. Victims, finding themselves in a new situation beyond their comprehension, are dependent upon government experts whose will and competency to master the situation they come increasingly to doubt.

A Dialectic of Double Binds

The result of this change in circumstances for victims is their virtual entrapment in a double bind. Namely, they learn that they are neither sufficiently at risk to warrant definitive action by government nor sufficiently free of risk to allow for a return to life as usual. On their part, the government officials working with them are also in an equally conflicting situation. They must respond to the public's concerns but without agreeing to take steps that extend beyond their regulatory authority, their budgets, professional norms, or political realities.

For both parties, a "double bind" results—one that is particularly evident in the context of communication between them. The term "double bind" originated from Bateson's efforts to explain schizophrenia by looking at family communication. Schizophrenia, in his view, results from contradictions between communications and metacommunications (communications about communications) that simultaneously send a message and negate that message. The result is endless interpretation that leads to "an experience of being punished precisely for being right in one's own view of the context" (1972, p. 236). Repeated experiences lead the parties to behave habitually as if they expected such punishment. A mutual double bind develops, such that "neither person can afford to receive or emit metacommunicative messages without distortion" (1972, p. 237).

I am not interested here in the long-term development of schizophrenia among individuals but rather in the prolonged dialectic that evolves from the inherently contradictory yet mutually bound relationship of regulator and victim of contamination. For the victim, there is an initial dependency on government officials. But when these officials simultaneously communicate such double messages as "You are at risk/ You are safe" and "I will help/ There is nothing I can or will do," victims' stress is exacerbated by these mixed messages. In turn, citizens demand help from regulatory officials, but reveal increasing degrees of frustration, anger, distrust, and hostility as they realize officials are not clarifying and solving problems, at least quickly. When regulators receive such mixed messages as "Do for me as much a you can/ It's not good enough" and "I am relying on your help/ I not only don't trust you but I blame you," they become embattled. The onset of this dysfunctional pattern of communication is illustrated by an episode that I observed during a meeting following the discovery of contamination in the Washington Heights section of Wallkill, New York in the mid-1980s.

It was a tense public meeting. Residents from the neighborhood crowded into the fire hall to learn more about their recently discovered contamination problem. The solvent tetrachloroethylene (PCE), a suspected carcinogen, had been found in wells at one end of the section. The concentration in one of the homes was among the highest ever recorded for a residential water source; a glass of tap water was said to be one-fourth PCE. Initial testing had found decreasing amounts of PCE as samples were taken at greater distances from the worst wells. In fact, other than at a cluster of some ten homes, no other wells had been found to have more than the 50-parts-per-billion (ppb) standard used by New York State to define the acceptable level of contamination; many wells showed no pollution at all.

Midway through the meeting, the county health commissioner arose to address the question of whether people should drink their water. Earlier, he had been quoted to the effect that water below the 50-ppb standard was safe to drink. He now sought to distance himself from this advice, noting that the aquifer beneath the neighborhood might never again be free of PCE. Residents would have to weigh the risks of continued use of their wells. As he followed this new line of reasoning, however, the commissioner was caught in a bind. If PCE was in the aquifer, it might show up in tap water at a later point. Therefore, the water could never again be trusted and should, accordingly, not be used. If the water could not be used, then something would have to be done to assist the residents.

Suddenly realizing the danger of overstressing this point, the commissioner attempted to balance his message with reassurances based upon the water standard. The message was now garbled, with residents being advised that their water was both safe and unsafe. At my urging, he attempted again to clarify this point, again unsuccessfully. Only later, as people were filing out the door, did he find the words to shout a clear warning, but few residents heard him. It was not surprising that some residents, having been told that their water supply was "safe" by government standards, subsequently returned to using their wells; the threat was over for them. Others, having heard that there was a potential ongoing threat, continued to believe that they were at risk. But their belief was supported neither by most government communications, nor by many of their neighbors. Denial of the threat was clearly an easier route than acceptance, in part because of the regulatory response. The dialectic of double binds had created a disabling situation.

The Citizens' Bind

Most Washington Heights residents were thrown into limbo because their wells were not polluted quite enough to trigger an arbitrary state action level. While their neighbors who had experienced the high levels of pollution were quickly provided with a new water system, these marginally affected residents were left with the uncertainty implicit in the health commissioner's remarks. Ironically, they soon discovered from PCE victims in Vermont that a lower level for state action had been adopted there. Had New York used the same level, assistance would have been triggered for many additional residents.[1]

Citizens' bind similarly characterized residents of Love Canal who lived just outside the arbitrary boundaries delineating "safe" from "unsafe" areas and who thus were deprived of tax benefits, offers of relocation, testing, and other government aid.[2] A further example involved Michigan farmers whose cattle were classified as having "low-level" rather than "high-level" PBB contamination in 1975, in the aftermath of the incident involving tainted feed. "Low-level" farmers received no compensation even when their herds showed severe symptoms of PBB poisoning resulting in economic losses. Ironically, some "high-level" herds were comparatively healthy yet qualified for assistance because of the higher PBB results (Reich, 1983; Coyer and Schwerin, 1981).

As these examples illustrate, government officials often deal with uncertainty by surrounding their decisions with an aura of seeming certainty. Once having made a decision, they are likely to use a scientific rationale to legitimize it, dismissing problems with their reasoning. Once locked into a position, the agency defends it stubbornly. Thus, what may begin as an open inquiry under the scientific method is transformed into a distorted approach under the guise of science. For example, officials attempted to define the Michigan PBB contamination as an agricultural problem rather than a health problem, a means of diffusing and bounding the crisis (Reich, 1983).

In other instances, as victims become alarmed by the potential dangers of health effects, a dominant concern in most incidents, they discover that health officials attempt to play down their fears.[3] Officials may be responding to legalistically defined rational criteria for categorizing different pollution cases. As a result, response is likely to be triggered only in cases of certain routes of contamination that are directly linked to health threats (such as water pollution), particularly if the resources and technology are at hand for addressing that type of problem and if housing is proximate (van Eijndhoven and Nieuwdrop, 1986). In determining the priority for response, triage is practiced with all but the most extreme contaminations. Thus, in Wallkill, residents having the extremely high levels of PCE at least had a clearly defined situation. The remaining residents were left in continuing uncertainty, their levels sufficiently low to not demand a regulatory response.

While the uncertainty of circumstance is most marked for the more marginally contaminated toxic victims, two issues must be kept in mind. First, nearly all toxic victims are somewhere within the marginal category. Second, most toxic victims suffer from citizens' bind. In seeking publicity, they enhance their community's stigma. In actively seeking answers, they enhance their level of stress. In depending upon government for assistance, they are likely to be disappointed. And facing a "mitigatory gap," wherein an extended period of time elapses between the definition of the exposure and the execution of steps to correct it, victims may find themselves trapped in a situation where they are damned no matter what they do. Even under the Superfund program in the United States, designed to expedite the cleanup of hazardous waste sites, remediations that have been done do not involve the level of cleanup that would restore residents' confidence in the security of their homes. As has also occurred in the Netherlands, government is likely to select the least expensive remediation approach rather than the level of restitution desired by residents (Melief, 1986).

The Regulators' Bind

Regulators' bind, in contrast, involves the frustrations of the local, state, and federal environmental and health officials. Acting as public servants seeking to solve a complex problem, they find themselves bound by regulations, political realities, limited resources, and their own perspective toward their task. Bureaucratic organizational structures create fragmented patterns of work and responsibility that result in dehumanizing practices and make likely overlapping, confusion, gaps in offered response, and poor coordination among different agencies working in the same community. Such organizations may also be subject to "regulatory capture" (see Reich, 1983; and Coyer and Schwerin, 1981) in the sense of having a built-in conflict of interest in favor of one side of a controversy, most often corporate, not citizen, interests.

In addition to these distorting influences, government officials face the complex task of discerning the fine line between danger and false alarm. In some cases, standards are set and must be applied. In other cases, no clear standards exist. Gray areas persist even when standards are available. For example, the health commissioner at the Wallkill meeting, in assessing the community threat, undoubtedly also had to weigh a host of considerations that are highly salient and concrete to a bureaucrat. His conflict was genuine. He was addressing an inherently gray question—how risky is exposure to small quantities of organic chemicals? He wanted neither to be responsible for failing to warn the people of the risk nor to be responsible for the consequences of a clear warning. Among the latter concerns were the fears of overcommitting his agency, overstepping his authority, causing "undue alarm or panic," becoming committed to the expenditure of unavailable funds, varying from the policy of the state health department, or taking on legal liability. He may have been uncomfortable pointing to the need for steps, such as the provision of a totally new water source, that he thought would be very difficult to achieve. His problem, in short, was how to simultaneously tell people that their water was unsafe and that he would do nothing about it—without their becoming understandably concerned.

The manner in which government officials communicate their evaluation of a situation creates the basis for the public's reaction. There may be no situation more feared by an official than the creation of panic. Thus, we can understand the pressure on the healh commissioner to play down risk, even in the face of perceived justification for concern.

Furthermore, in a toxic incident, somebody has to determine the operational standards for risk, and whoever does is likely to take the brunt of criticism. Given the large number of unknowns in a toxic exposure incident, it is a thankless task to be responsible for bringing clarity and certainty to the situation. With the announcement of exposure, citizens look to government experts to inform them and to fix the problem. For an official to guess at the answers demanded may be rash, to hold back cautious. But at some point, citizens unappreciative of caution expect officials to have the answers. Invariably, this places the regulators in a no-win situation. Reich (1983, p. 303) cites a former EPA official, Steven Jellinek, on this point. "The regulator must make decisions about chemicals 'in the midst of pervasive uncertainty.' Since the regulator does not have 'the luxury' of putting off decisions until certainty arrives, there exists an 'inevitability of being wrong' sometimes."

Reko describes the consequences of this probability of error for some regulators at Times Beach (1984, p. 40). "As the residents pressured the EPA with questions and demands for solutions to their problems, the administrators felt caught in a bind. 'I could not give straight answers because there was no solution. We were taking the blame for everything that had, or had not, been done, and it was unfair,' complained one local EPA representative."

The Consequences of Citizens' and Regulators' Binds

Citizens' and regulators' binds interact synergistically. These binds make it highly unlikely that alarmed citizens will understand and accept regulators' decisions as rational. As a result, the very people the regulators attempt to help increasingly vilify them, making officials the scapegoats in situations that they didn't create. Any action these officials take is likely to be "wrong" according to one of their constituencies. The effects of regulators' bind are suggested by Harris (1984, p. 428) in these comments about health officials:

> In their battles for better sanitation, decent housing, milk pasteurization, and maternal and child health services, health departments have generally enjoyed the support of citizen reform groups. To find themselves now labeled by environmental activists of the 1980s as the enemy is a stunning reversal of history and a shattering blow to their self-perception as the champions of the public interest.

As a result, much as toxic victims are stressed from their interaction with regulators, the officials are also highly stressed from these encounters. Thus, as a therapist, Tester (1982) treated a number of regulators who were greatly distressed by their inability to respond effectively. Similarly, a government official confided in me about the fears he had on several occasions. At one point, he feared reprisal by organized crime if he was too stringent in regulating a landfill. At another time, he seriously expected to be lynched when an angry crowd of farmers could not be controlled by their elected officials. Levine (1982) sheds further light on the pressures faced by regulators in her discussion of the efforts of officials to maintain a professional manner in the face of citizen anger.

> The person who conducted the meeting on that cold, gloomy morning in December mentioned later that he felt proud he had been able to remain cool, "to talk like a machine," despite the anger of the Love Canal residents displayed when he read the DOH announcement. Task force members expressed similar feelings more than once when they conducted or even were present at meetings with residents who seemed so unreasonable. Privately, the officials congratulated each other on not giving anything away, on not conceding anything to the residents. In public, they said they saw "no evidence," or said "I don't know," and argued that the people were given answers, but "like spoiled children," just did not like the answers they received (p. 98).

That regulators at times become resentful of such pressures was evident at a statewide conference for toxic victims that I attended. A New York

State health official verbally attacked the audience until they responded to him in an equally hostile way. This official unconsciously played out the dialectic with citizens that characterizes their mutual binds. Rather than creating a better understanding through the exercise, however, the interaction produced even greater mutual animosity.

What are the major sources for the disabling dialectic between toxic victims and regulators? Beyond the fact that their basic interests vary, it is evident that there is a flawed process of communication between these groups. There is additionally a divergence in the paradigm for defining acceptable risk that brings about a conflict of communicational content. These two contributions to disablement will be examined.

Communicational Distortion in the Institutional Context

A major source of stress and controversy is the form taken by regulatory communication. As threatened communities attempt to cope with exposure, information becomes a vital commodity. This is particularly true when the situation is novel and there is great uncertainty experienced by individuals and communities alike. However, government agencies tend to maintain strict control over information (Paigen, 1982). This strategy reinforces both the initial dependence and helplessness of citizens as well as their eventual distrust of the agencies.

Furthermore, regulatory communication tends to vary from the expectations for valid communication used in everyday life as judged by four criteria: the clarity of the communication, its basis in fact, its believability, and its appropriateness. The failure of either or both parties to meet even one of the four criteria may result in the parties breaking off communication, arguing over unresolved issues, or adopting a mode of action aimed at forcing the issue in some way (Habermas, 1979). All too often, regulators fail to communicate clearly, their assumptions are questioned, they come not to be trusted, or their responses fall short of citizens' expectations for what is called for by the situation (Edelstein, 1986/1987; Forester, 1980). Partially as a result, citizens often use attention-getting strategies, inappropriate action by the standards of many officials. Citizens may also be unclear, inaccurate, or unbelievable.

Given their lack of preparation for the situation, it is understandable that citizens may have difficulty communicating effectively. Government officials should not have the same excuse. When institutional communication to toxic victims fails to meet these criteria, the result is distorted communication, defined by Mueller (1973, p. 19) as "all forms of restricted and prejudiced communication that by their nature inhibit a full discussion of problems, issues and ideas that have public relevance." He delineates three types of distorted communication.

1. Directed communication occurs when government policy structures language and communication. For example, in toxic incidents, regulations guide the definition of "the problem," the rights of victims, and the responsibilities of agencies to respond. The language of regulation comes to dominate much of the exchange. Furthermore, this language relies heavily upon the vocabulary and methods of science and engineering.

2. Arrested communication occurs when some groups cannot engage in political communication because they lack the linguistic ability to do so. For example, the average citizen lacks the technical expertise to participate in many of the specialized decisions that follow from toxic exposure. Accordingly, Harris (1984), a government official, suggests that "scientific illiteracy" is a basic source of distortion in regulator/citizen interactions. If citizens had a better grasp of the scientific and technical issues involved, he argues, they would be better able to understand the reasoning of the regulatory experts.

A contradictory view, suggested by the leaders of community groups, is that citizens are capable of mastering complex environmental concepts, but that government agencies impede their acquisition of information (Freudenberg, 1984a). The use of expert consultants by citizens may help to compensate for this imbalance in background (see Chapter 6). Additionally, it can be argued that government officials need to improve their own "arrested" skills in communicating with the public.

3. Constrained communication involves actions by private or governmental groups to seek their own self-interest by structuring and limiting public communication. This form of distortion invites distrust more than any other. Examples of constrained communication by government officials include giving false reassurances, withholding information, responding slowly, taking inadequate actions, mishandling testing, arguing among experts, taking actions secretively, drawing arbitrary boundaries, using insensitive or incomprehensible language, denying citizens' perceptions and fears, hiding behind scientific reasoning for political decisions, and appearing to not know what is going on. There are times, of course, when these techniques are used unintentionally. But they are also used strategically at times to distort an issue. Levine (1982) provides a clear example of such strategic distortion in her description of the "Thomas Commission," the panel of scientists that reviewed findings about the Love Canal for the New York State Department of Health.

The distortion of regulatory communication has strong roots in agencies suffering from an inherent contradiction—that they are nonindependent scientific entities. Thus, they combine severe institutional limits on inquiry with a method that is based on the objective search for truth. Similarly, government officials experience conflict in filling such roles as "public servant," "bureaucrat," "scientist," "technocrat," "promoter of private enterprise," and "regulator of compliance with public laws." Their performance is further complicated by the politicized nature of their work context (see Levine, 1982).

Furthermore, distortion is enhanced by the situation, as illustrated by a generic communication problem—how to present test data to toxic victims. For example, some time after the new central water system in Legler restored some semblance of normalcy for the residents, a housewife opened her mail to discover the results of air quality tests in her basement. An impersonal letter from the New Jersey Department of Environmental Protection implied that a dangerous situation existed, but the letter offered little information by which the family might assess the risk. This communication had a major emotional impact upon the wife. The husband recalled, "When I came home, I found my wife shaking; she was drinking." The wife added,

> *I couldn't calm down. Look what was in there! [pushing a list of chemicals toward me] Reading this and thinking about it, it bothers me as much as the water problem, almost more. We don't know what's going on. We understand that these are dangerous chemicals, but not how dangerous they are. We want to talk to somebody who isn't involved.*

Toxic victims frequently note that a source of information independent of government is needed if trusted interpretations are to be made. As it was, this family was left in a panic situation. They worried for weeks about the results without making any attempt to seek clarification. And yet, when I read the state's letter, I discovered that the couple had misread some of their results. I also noticed immediately that there was a standard invitation to call if there were any questions. But the overriding message of the letter so effectively made it seem like a one-way communication that the couple had not even considered calling. In the wake of the water contamination incident, they had also obviously ruled out government as a trusted source of information. Barriers to effective communication were thus evident both with the receiver and the sender.

Another example of miscommunication regarding test results is found in the response to several dioxin incidents by the Missouri State Division of Health (Miller, 1984). When the department undertook a study of the incidents, the atmosphere was so highly charged that the study directors were under great pressure. As a result, they left the public with inaccurate expectations both about how quickly the study could be completed and about how definitive it would be in identifying health effects from dioxin contamination. This set up a later crisis when, with the study incomplete, the agency was forced to release preliminary results. When the study group members met to decide which results to release, they used two considerations. First, they wanted to avoid later accusations that data had been withheld. Second, like health officers working at Love Canal (see Fowlkes and Miller, 1982), they believed that the family physician should be involved in interpreting the significance of a test result for a given individual. As a result of these decisions, the data released included all findings exceeding the norm as well as all abnormalities identified, even those which the researchers knew had nothing to do with exposure to dioxin.

When residents received these results in the mail, they were highly alarmed because the simplified summaries had exaggerated the extent of adverse health findings by listing all abnormalities. The manner in which results were summarized added to the confusion. Hoping to simplify matters, the agency had used general statements (e.g., "abnormal blood chemistry") rather than specific results (e.g., "elevated cholesterol [293]"). In order to identify the actual results, recipients had to examine a complex laboratory report. While the officials had actually sought to provide full disclosure, residents felt either deceived or patronized by the simplified information. Furthermore, the assumption that residents had access to a physician who would or could help them interpret the results did not prove to be true. Thus, being told that they were abnormal on what they had assumed to be a test for dioxin contamination was enough to thoroughly alarm most recipients (Miller, 1984). Distorted communication in this instance involved the violation of appropriateness, clarity, truthfulness, and authenticity, despite the intentions of the agency to be open about all possible findings.

Another example of miscommunication in Missouri occurred when the EPA varied from the mitigation strategy used at previous local dioxin sites. Rather than evacuating residents at Castlewood, officials decided to leave residents in their homes during remediation. Although to the EPA this was justified by the relatively low levels of contamination and the delays foreseeable in any attempt at permanent relocation, citizens viewed this new procedure differently. The EPA took great pains to describe its planned remediation, but made little attempt to discuss the reasons for selecting this approach. The result was to make "a perfectly reasonable decision seem arbitrary and capricious" (Miller, 1984, p. 22). This led to what Miller describes as "a series of acutely frustrating encounters in which residents ask the wrong questions and officials respond with the wrong answers" (Miller, 1984, p. 22). Thus, in one case, residents of a street not included in a sampling program asked why they weren't being tested. Instead of giving detailed reasoning for the EPA's belief that there was no contamination there due to soil conditions, the official response was merely to note that the street hadn't been selected for sampling (Miller, 1984). Again, the norms for complete communication had been violated.

Differing Paradigms of Risk
Between Citizens and Regulators

Controversy is inherent in the relationship of toxic victims to their institutional context because there are differences in the way citizens and technocrats view risk. The "technocratic paradigm" of risk uses a scientific and technical rationale in making decisions for citizens (Dickson, 1981). Underlying this approach is the desire to protect the business community from adverse impacts, a desire that is basic to the modern concept of regulation (see Polanyi, 1944). The technocratic paradigm thus creates regulations based on risk estimates that equate "caution" with steps that least threaten the private sector. The significance of claims by the victims of corporate pollution must be certified by experts to have legitimacy. For controlling risks, the marketplace is preferred to regulation (Dickson, 1981).

The contrasting "democratic paradigm" takes the victims' perspective in estimating the probability of technological hazards. "Safety" is valued over profit; prevention is triggered by less than conclusive proof. Most importantly, those bearing the impacts are given an active voice in determining acceptable risk and in making decisions (Dickson, 1981).

Given the lack of models for citizen participation in decisions resulting from toxic exposure (Shaw and Milbrath, 1983), it is not surprising that regulatory agencies proceed with the problem-solving approach characteristic of the technocratic paradigm. In contrast, the democratic model is readily adopted by toxic victims. These paradigms are in direct opposition. It is no wonder that fact and value are so confounded during technical controversy (Mazur, 1981). Not surprisingly, there are major role differences between victim and regulator as well as differences in the ways that acceptable risk is defined.

The Role: Professional Versus Victim

The difference between the psychological environments of citizen and regulator is captured in two examples. First, Levine (1982, p. 36) reports of a Love Canal incident that "The state health officials thought they were offering help in this dreadful situation in a way that was reasonable, concerned, and within the department's legitimate purview and capabilities. In the words of a young man pointing to his pregnant wife, however, 'The damage is done! My wife is eight months pregnant. What are you going to do for my baby? It's too late for my child.'"

And during the Chemical Control fire in Elizabeth, New Jersey, when thousands of barrels of toxic materials exploded, the Relocated Bayway area was totally blocked off, trapping residents. As one resident described the incident to me:

> *Nobody came and gave us a warning. The only way you can get out is to swim across that river. How's you going to get out when the streets was closed? Nobody could come in and out. And then finally one guy comes with his car, real fast. EPA man. He jumps out of his car and opens his trunk. And he's putting his mask on. I says, "Hey, where the hell is ours?" And he had everything for himself. Well, where's our stuff? And they told us not to go outside. And then you know what? Your house was closed, your windows was closed. Do you know, the bedding, the sheets and the pillowcases was burning your face that night. How did it get into the house? All that—your face was burning! If you wet with your tongue, your lips were burning.*

As both of these examples suggest, for vigilant victims of toxic exposure, the threat is near. It is in their homes, their neighborhood, their air, their water, their bodies, and their children's bodies. This proximity makes the threat "real." It may be invisible, but it is pervasive. They see it everywhere. It is no longer just part of their physical environment; it has entered their perceptual environment as well. It is not just out there in an out-of-sight-out-of-mind way. It is *here!*

In contrast, for government officials or experts, the threat is a faraway abstraction. While they may be responsible for averting danger, they can be content to do so in an "objective way"—after all, they are not personally involved, so why should it be a subjective experience? They may visit the site for inspections, suitably protected, but then they leave. At home that night, they can put work from their mind. Given their technical and scientific training, they may easily intellectualize the problem. In fact, by treating it as a *problem*, it has already been conceptually packaged for them. Unlike the victim who feels a silent killer is stalking families, the regulator can see the issue as a work problem that must be solved to the greatest degree possible given available resources.

Accordingly, the differences in the immediacy and meaning of the experience for officials and citizens are at the root of the communications gap between them. While officials are professionals attempting to do a job according to the guidelines describing their responsibility, citizens are often in the role of victim, responsible for the well-being of families and threatened by newly discovered facets of the environment. In short, the two groups approach their encounter from quite different perspectives.

Perhaps these differences surface most clearly in answering the question of what risk is acceptable for a given community. A rule of thumb (not a universal) is that a given hazard is most acceptable to those farthest away and thus least vulnerable. In most cases, the regulators are more distanced than the residents—not only by their professional role but also by the lack of proximity of their homes and degree of exposure to the site.

Conflicting Definitions of Acceptable Risk

Through my observation of the interpretative biases of citizens versus those of government and industry experts, an analogy can be drawn to the statistical concept of the Type I and Type II error.[4] Put simply, for the scientist a Type I error is "an error of rashness," when one concludes that an effect occurred when in fact it did not. Conversely, a Type II error is "an error of caution," when one disregards a real effect because of insufficient proof (see Scott and Wertheimer, 1962, pp. 204–205). The norms of science attempt to guard against spurious conclusions that erroneously support theory; a theory that is truly predictive will be able to survive a stringent test. Accordingly, scientists strongly bias their work in favor of committing Type II errors and avoiding Type I errors. In the context of toxic exposure, a conservative judgment for a government or industry expert means caution in concluding that a place is *unsafe* (see also Levine, 1982; and Paigen, 1982).

Because a determination of significant contamination may result in demands for costly government action, reliance on scientific purism can serve as a convenient excuse for avoiding action. At Love Canal, for example, regulatory interpretation of risk was clearly affected by the concern over who would pay to remedy the hazard.[5] Such pressures incline an agency to favor "optimistic interpretations" (van Eijndhoven and Nieuwdorp, 1986). The politically (and economically) motivated use of conservative risk has more recently characterized the Reagan administration's approach to toxic hazards (see, for example, Lasch et al., 1984; Marshall, 1982).

In contrast to such "Type II conservatism," potential or established victims of toxic exposure are likely to evidence a "Type I conservatism." From the concerned citizens' perspective, it is caution, not rashness, to risk an error in defining a place as dangerous; overreaction to a potential threat is generally preferable to underreaction. Citizens tend to prefer "worst case analysis" to the "best case analysis" of the regulators (van Eijndhoven, et al., 1985, p. 6).

These differing views about risk reflect opposite normative frameworks for defining an "appropriate" response; they also reflect contrasting perceptions of truth. A perfect example of the incompatibility of these views is seen in the use of the word "significant." To a scientist, a significant threat is one in which a statistically improbable trend or cause-and-effect relationship is shown to occur. Normally, such findings must demonstrate less than one chance in twenty of occurring randomly to be termed significant. Of course, this basis for judgment is rarely explained to toxic victims, who may read or hear comments about "significance" and "nonsignificance" without understanding why the terms were chosen. The distortion involved in using these terms can be seen when they are applied to something of keen interest, such as a person's water test results. What is not significant to the regulator may be very much so to the resident. These differences in defining conservative risk are a major source of conflict over how to address contamination, as Levine found at Love Canal (1982, p. 66).

> Most residents wanted to leave open the definition of how much harm had been done to their environment and to their health. They wanted the decisions about essential help left flexible, so that appropriate help could be offered to fulfill needs as they became apparent. The officials wanted the definition of harm to be narrow, to fit the resources they had available, so that some tasks could be successfully accomplished, accounted for, and pointed to. In a situation with so many unknowns, the people wanted to feel that the available help would be sufficient to meet their needs, even if some of their needs could not yet be anticipated; the officials wanted to define some aspects as *known* and to address their attention to these. Just as they were to fence in a defined area around a construction site when the underlying chemical leachates might be flowing far beyond, so they wanted to define and bound the problem to be solved, even though it lay within the problem that in fact *existed.*

The resulting risk dynamic is likely to lead to a perceived mutual loss of trustfulness on the part of both regulators and citizens. The citizen is viewed as a "screamer" by the regulator for reacting to risk that is not scientifically established; the regulator is seen as "corrupt or incompetent" by the citizen for denying the potential for risk. In a sense, the citizen views the chemical as guilty until proven innocent (as Levine, 1982, points out), while the official sees the chemical as innocent until proven guilty (see Brown, 1980).

The inverse of acceptable risk involves definitions of adequate safety. Here again, citizens and officials are likely to differ. Thus, unlike citizens exposed to PCBs who are concerned with the potential effects of the "absolute" exposure, officials are more likely to weigh the "relative" risks of differing health threats in judging how much attention to pay to a particular PCB exposure. The regulator is willing to allow "reasonable acceptable risk," while the citizen seeks to prevent all "unnecessary risk."[6]

The relative-risk approach is illustrated by a report issued by the New York State Department of Health (Kim and Stone, 1980), which informs the reader that the risk of death from drinking two liters of water per day contaminated with organic chemicals is relatively small. For example, the report projects that this exposure to tetrachloroethylene (PCE) would produce 22 cancer deaths for every 100,000 people exposed for a 70 year lifetime, while PCBs would cause 1.6 cancer deaths under similar conditions. In contrast, the lifetime risk of death due to motor vehicle accidents is an impressive 1,750 per 100,000, and the risk of respiratory cancer alone for someone smoking just one pack of cigarettes per day is an even more dramatic 4,000 in 100,000.

Given such comparisons, a public health officer can hardly be blamed for holding the view that smoking and driving are much greater public threats than is consumption of minute quantities of PCB or PCE. According to scientific definitions of significance, the priorities are clear, as noted by Harris.

> It would come as no surprise were public health officials to report high levels of frustration as they encounter relentless pressures for the commitment of scarce resources to combat ill-defined environmental hazards, but little popular enthusiasm for taking aim against the proven and terrible hazards of smoking, alcoholism, poor diet, lack of exercise, and failure to simply buckle up in the family car (1984, p. 430).

The victims of exposure are likely to reject and resent setting priorities according to risk estimations, at least in a specific exposure context. Due to the involuntary, unnatural and intrusive, and unknown and unseen character of toxic exposure, the victim is unlikely to appreciate statistically weighted risk comparisons that minimize the significance of toxic threat.[7] Thus, when my co-chair at a public meeting called attention to the fact that toxic victims in the audience who were smoking were accepting a greater risk than that entailed in drinking their polluted water, they totally rejected her argument, seeing no relationship between what they chose to do and the contamination that was forced upon them. More than a year later, they were still commenting about "the nerve" my colleague had in raising this issue.

Accordingly, what is a relative risk to experts may appear as an absolute risk to citizens. It does not matter that the same citizens overlook (through denial, rationalization, or other distortions) an even greater threat from a source presumably under their control. Conversely, what is a significant threat to officials may be only a relative threat to citizens.

A final twist on this question of absolute and relative risk involves the existence of standards setting maximum levels for some chemicals in the water, air, soil, or other media. The existence of these action levels forces officials to treat contamination as though there were a threshold for absolute concern. As a result, regulators discount as "insignificant" levels of contamination below the standards. Yet this is a purely legalistic response and does not reflect the minimal support for the scientific basis behind these standards. Thus, citizens in the Washington Heights section of Wallkill, New York, legitimately asked why PCE contamination in their water had to reach 50 parts per billion (ppb) to be a significant risk when citizens of Vermont and Washington only needed 30 ppb to trigger government assistance (see also Miller, 1984).

Thus, the question of absolute and relative risk is reversed when standards exist. The victim, who applies an absolute criterion in judging the significance of contamination, uses a relative standard in assessing how much contamination is acceptable. The official, who sees risk as relative, uses absolute standards to define when contamination demands action. These definitions of risk serve as the basis for the respective double binds of citizen and regulator.

Institutional Contexts

The key institutional contexts associated with disabling dynamics vary for individual toxic incidents, but the dynamics are inherently similar from incident to incident. Although different levels of government are involved in every incident, one usually predominates. In Legler, the township was the key actor; for the Missouri dioxin sites, federal agencies played a central role; at Love Canal, the state was the main responder. The Legler case has already been discussed at length. Following is a brief review of the latter two contexts.

Times Beach—The Feds as Disablers

In the Times Beach dioxin contamination case, where federal agencies predominated, a major conflict occurred between the Centers for Disease Control and the Environmental Protection Agency. The CDC forced risk levels to be set at one part per billion. The EPA unsuccessfully attempted to raise the level to one hundred parts per billion, thereby reducing the number of "significant" instances of contamination by half (Miller, 1984). Clearly this debate had major ramifications for local citizens, but they were shut out of the discussion. They were further excluded when then-EPA director Ann Burford announced the long-awaited Times Beach buyout, addressing reporters face-to-face in a local hotel while residents were forced to listen to her speech through a glass wall that separated them (Reko, 1984).

Love Canal—The State as Disabler

Among the many actors at Love Canal, New York State (particularly the Department of Health) played a particularly disabling role. Several examples from Levine (1982) show what citizens faced in this institutional context. The commissioner of health's dramatic August 1978 declaration of a health emergency, which advised evacuation of pregnant women and children under age two, created confusion from the onset. The order failed to clarify what was to happen to the rest of the family, who would pay for the evacuation, and whether others remaining were safe. That officials never even considered such clarification to be important is evidenced by an official's comments to Levine (1982, p. 40).

> The health department professionals were *scientists*, who did not worry about people's reactions to cautionary statements and recommended actions. They dealt with numbers—with data on physical conditions—and only with these. Political and social matters, the official stressed, were extraneous to the DOH work. The issues of how pregnant women and children would move to safe places, how people might feel about recommendations from the state health commissioner not to go into their basements and not to eat food grown in their gardens, were not seen as the responsibility of scientists.

Instead of dealing with people, the state focused on the technical plan to mitigate the threat. The plan, considered by residents as inadequate, was promoted by both officials and state politicians concerned less with a health threat than with a political threat—they needed to attract federal funding for an expensive project while simultaneously minimizing expensive precedents. As federal guidelines for funding changed, the state redefined the situation, downplaying the health threat and emphasizing the need to move people in order to conduct the remedial construction, which involved covering the landfill in order to limit the flow of leachate from the site.

Meanwhile, Love Canal residents were expected to feel safe when their neighborhood was separated from the condemned area along the canal by a fence and warning signs. Such precautions did not keep children from playing in the area; in fact, a supervised play program was held during the summer of 1978 on the school playground, which sat directly on the canal.

The faith of residents was further shaken by open disagreements among government representatives, clearly unwarranted reassurances by officials, jargonistic test results not easily translated into common terms, and the collection by the state health department of so many blood samples that it could not process them all. To residents, the state appeared to be hiding behind excuses that more research was needed before health effects might be determined. Furthermore, the state health department denigrated citizen-collected health data as well as the reputation of the scientist who analyzed the data. Additionally, it initially dismissed the presence of dioxin in the neighborhood and handled findings of liver damage among children in a manner suspicious to the residents.

While residents were excluded from the decision making, unnamed experts met at a far-off airport to advise the state. Later, the Thomas Commission, a scientific panel appointed by the governor, was additionally disabling because of its secretive control of information, its exclusion of residents from its discussions, and because its report was shaped by the governor's aides into a political tool for minimizing the problem. Rather than engendering the public's trust, the commission generated additional questions regarding political interference in a pseudoscientific process.

The result of the Thomas Commission findings, widely cited by the chemical industry as well as by government, was what Levine (1982, p. 168) terms a "semiofficial contemporary legend" about Love Canal. Proven health effects from the canal were discounted while emphasis was placed upon psychological damage due to the ineffective actions of officials, scientists, and the media and the influence of anticorporate forces, environmentalists, and emotional housewives. Accordingly, relocation occurred out of compassion or for the purpose of remedial construction, not because of a threat to health. In constructing this legend, state officials emphasized the uniqueness of Love Canal, hoping to rationalize the millions of dollars spent while minimizing the setting of precedents for other sites.

Conclusion and Summary

This chapter has focused mainly on the interaction of citizens and government professionals. I have generally avoided addressing experts working for private consulting firms or directly for hazard-related companies. While the main arguments made here certainly pertain to the response of such experts, the fact that they are seen by citizens as biased in favor of their employers is less disabling than is the realization that professionals working for government are biased. Few citizens expect industry experts to come to the rescue of toxic victims. Their expectations are quite different for government officials.

While one might envision enlightened government officials who overcome the constraints that bring about the disabling of citizens, the likelihood remains that this dynamic will continue to accompany toxic incidents in the foreseeable future. The scientists and engineers who fill many of the key environmental agency posts are trained neither in communication skills nor sensitivity to their nuances. Instead, they are trained to break the physical aspects of the situation down into "problems," but not to reconstruct these steps into an overall composite of the victims' experience.

Rather than resulting in collaboration between government and citizens to deal with the crisis, communication between these parties increasingly distances them. They become caught up in controversy over differing perceptions of the situation. The clash of risk paradigms appears to solidify each party's belief system. The kinds of communicational devices that promote trust are absent from these interactions. The victims now bear the weight of disappointment and abandonment by government in addition to their concerns stemming directly from exposure. The resulting frustration creates a strong pressure by victims to break loose from passive dependency upon government. It is possibly this reaction that is the strongest response to the institutional context of toxic exposure. Thus, as Reich notes of farmers responding to the Michigan PBB contamination (1983, p. 309), "The PBB case reminds us that to influence public policy, individuals and groups must transform a private trouble into a public issue and often into a public controversy." The next chapter discusses the process of enabling that accompanies this controversy.

Notes

1. Over the long term, residents in the community are being given the opportunity to join a newly formed water district at their own expense.

2. See Shaw and Milbrath, 1983; Fowlkes and Miller, 1982; L. Gibbs, 1982a; and Levine, 1982.

3. See Creen, 1984; Levine, 1982; Paigen, 1982; also Freudenberg, 1984a.

4. The analogy of Type I and Type II error was used independently by Edelstein, 1982; and Levine, 1982. It is also employed by Paigen, 1982.

5. See Fowlkes and Miller, 1982; L. Gibbs, 1982a; and Levine, 1982.

6. Drawn from an unpublished paper by Daniel Wartenberg and Theodore Goldfarb, p. 2.

7. See Vyner, 1984; Edelstein, 1982; and Levine, 1982.

6

The Enabling Response—
Community Development
and Toxic Exposure

In the introductory chapter, I defined a "contaminated community" as a residential area located within the identified boundaries for a known exposure to some form of pollution. In this chapter, I examine the social dynamics within such communities.[1]

Enablement Through Community Development

Community is often thought of as one of the victims of disaster.[2] However, in examining the effects of chronic, low-level contamination on many neighborhoods, one frequently finds the development, rather than destruction, of community.[3] In this context, community development refers to the creation of grass roots groups that are formed to represent the affected region. The commonality of this outcome is indicated by a study of 21 toxic sites (ICF, 1981, p. 34) that found "*ad hoc* groups were formed, often quite rapidly, at every site studied with significant public participation."

The previous chapter described how government response operates to *disable* individuals and families in their attempts to deal with the complexities of exposure. This chapter discusses how the development of community organizations serves to *enable* many toxic victims. It also examines the conditions in which community development is absent. Finally, it discusses the societal implications resulting from the networking of local groups into an impressive national social movement that is addressing concerns about toxic exposure.

As illustrated in Figure 6.1, grass roots participation develops because of a combination of occurrences that form a basis for at least temporary community consensus and cohesiveness: victims' normal lives are severely disrupted by the exposure incident, victims are isolated from their normal relational and institutional networks, individual families cannot solve their problems alone, and a group of proximate victims shares the same conditions.

In a society based upon the autonomous nuclear family within an independent home, the family initially faces the news of contamination as a private challenge or problem. But it is a complex and confusing situation beyond the family's resources to control. Consequently, in their attempts to cope with this disruption, residents of affected areas (newly labeled as "victims") turn naturally to key components of their social environments, the "social network" (the existing group of friends and relatives who are expected to offer support in the face of crisis) and the "institutional network" (government agencies believed to be responsible for providing clarification and assistance to citizens in need).

Support from one's social network is commonly identified as essential in attempts to cope with a variety of stressful situations (e.g., Sowder, 1985; Gottlieb, 1981). However, as previously discussed, networks based upon ties to relatives, friends, and co-workers outside the area of contamination frequently fail to support victims adequately. Similarly, when victims turn to government, disappointment is the norm.

Because neither the relational nor institutional network is likely to be of much help, toxic exposure results in a virtual isolation of victims from outside sources of support. As toxic victims discover that their previous channels for participation and problem solving are ineffective, they are forced to improvise new alternatives. The result of this isolation from key support networks combined with the geographic proximity of victims is the creation of a "spatial network," a community group corresponding to the boundaries of contamination. This grass roots organization appears to play a key role in attempts to cope with the demands of toxic exposure for individuals, families, and the newly defined community as well.[4] As a result, toxic victims have a common identity that has a significance beyond preexisting political, geographic, or social boundaries. They develop a "sense of community," which implies feelings of similarity, interdependence maintained by mutual support, and the sense that they are part of a structure that is larger and more stable than the individual can ever have in isolation (Sarason, 1974).

Signposts of this sense of community are readily found in communities confronting toxic exposure. For example, my fieldwork notes and newsletter files reveal this sampling of names for grass roots organizations: HARP (Heights Area Residents Against Pollution), CHOKE (Citizens Holding out for a Clean Environment), OUCH (Opposing Unnecessary Chemical Hazards), NOPE (Northern Ohioans Protecting the Environment), TWIG (Toxic Waste Investigative Group), DECALE (Deptford Citizens Against Landfills and Extensions), TRAPP (Tauton Residents

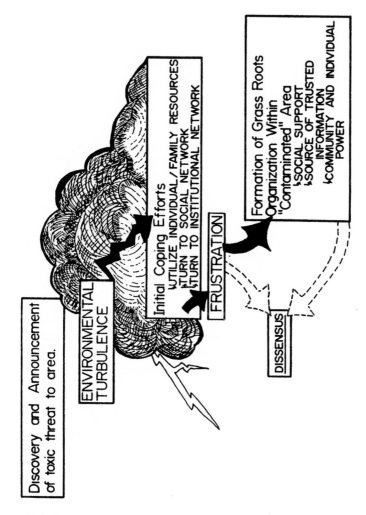

Figure 6.1 An Exploratory Map of Community Response to Toxic Contamination (taken from Michael R. Edelstein and Abraham Wandersman, "Community Dynamics in Coping with Toxic Exposure," in Irwin Altman and Abraham Wandersman (eds.), *Neighborhood and Community Environments* [New York: Plenum Press, 1987] and reprinted by permission of Plenum Press)

Against Possible Pollution), SOCM (Save Our Cumberland Mountains), WARD (Warwick Against Radium Dumping), and scores of "concerned citizens" groups.

These names display some of the flavor found in the community development that arises from residential toxic hazards. Such groups serve as the basis for the enabling response to toxic exposure. They represent collective coping mechanisms that provide for local control through the participation of commonly affected residents.

Keys to Enablement: Leadership and Activism

The availability of talented leadership with time to spend is a necessary prerequisite to organizing (ICF, 1981). A frequent scenario among toxic response organizations in the United States involves the development of a relatively stable core of leaders who carry the organization. For a consensual group to form and operate, it must also be able to draw upon a sufficient body of activist members who augment the leadership.

Who are these activists? Because of the diversity of communities facing pollution problems, participants cross the demographic spectrum. While specific characteristics of activists differ from location to location, a national survey characterized typical members as between 26 to 40 years old and homemakers. The leaders often had professional careers yet represented a diverse socioeconomic spectrum (Freudenberg, 1984b).

Women tend to predominate in leadership positions of groups concerned with toxic exposure (Shaw and Milbrath, 1983; ICF, 1981). This phenomenon may be due to a number of factors (Shaw and Milbrath, 1983; L. Gibbs, 1982b). In the communities that tend to suffer most from toxic contamination—blue-collar, lower middle-class, or poor areas—women tend to stay home and care for the children. They have established social networks available for quick action, and they are more likely to recognize patterns of ill health in the neighborhood. Men in the affected communities often work for polluting industries. They may see themselves as having failed in their role as protector of the home, or they may view as unmasculine the public admission of fears and concerns over health and safety. Finally, unlike men, women may be a little more distanced from the dominant social paradigm of economic growth and, therefore, less willing to rationalize risks as necessary for economic well-being (see Milbrath, 1984).

An in-depth picture of activists at Love Canal based on Levine's intensive interview study sheds further light on community involvement. Activists were most often found to have new roots in the neighborhood; they were neither old-timers nor new arrivals, but had lived there long enough to develop an attachment to place (typically six to ten years).

They tended to come from families with one wage earner (not a chemical worker) and young children, to own their homes, to have at least moderate incomes, and to have better educations and smaller families than did nonactivists. Activists tended to be under age 40 (Stone and Levine, 1985). A similar profile is suggested by a review of groups at other sites (see ICF, 1981).

Key Benefits of Community Development

When a shared sense and practice of community occurs, a community group can serve as a "therapeutic social system" for a very stressed neighborhood. Such therapeutic communities help compensate for the stress of disasters by sharing information, warmth, and help; providing a means for rapid consensus about actions necessary to meet collective needs; and engendering a high degree of motivation among victims to work for common purposes (Barton, 1969; see also Sowder, 1985). There are three key outcomes from the grass roots response to toxic exposure.[5] These outcomes—social support, information, and power—will be briefly reviewed in light of various case studies.

Social Support

Social support provides reassurance and strength in the face of a shared predicament (Schacter, 1959), helps disaster victims avoid the feeling of being abandoned, may help to increase their accuracy in estimating danger (Wolfenstein, 1957), and is a key to successful coping with technological disaster.[6] Social support may be task-oriented or focused upon personal/emotional needs; its concern may be at the individual or neighborhood levels.[7] As toxic victims organize, they build not only an organization but relationships as well. In the absence of either external validation or correction of their fears, victims are forced to turn to each other (Creen, 1984). Their need for social support as a result of the isolation of their situation (see also Baas, 1986) is met, in part, by turning to their neighbors.

Legler, as noted, was a "bedroom community" where social interaction within the neighborhood had been limited. But after the water problem, residents participated in the community organization, developed a sense of community, and engaged in "neighboring," defined by Unger and Wandersman (1985, p. 141) as "the social interaction, the symbolic interaction, and the attachment of individuals with the people living around them and the place in which they live." The Concerned Citizens Committee helped people get to meet others with whom they shared what otherwise seemed to be a private experience, inaccessible to the understanding of nonvictims. People were reassured that they were neither crazy nor alone (Edelstein, 1982).

Similarly, Levine reflects on the Love Canal Homeowners Association (1982, pp. 185–186; see also L. Gibbs, 1982b and 1985):

> As time passed and discouraging incidents piled one on the other, residents felt more and more that no one but people undergoing the experience themselves . . . could understand them. Only fellow sufferers could share the feelings of uncertainty . . . . As they confided in each other more and more, both privately and in public; as they worked together and participated in individual and group actions . . . ; as they interpreted their beliefs, their goals, their behavior; and as they thought and planned together for a better and happier future, free of Love Canal, they became essential supports for each other—a sort of large self-help group, a substitute for a strong, supportive, and ever-understanding family.

Interestingly, while Love Canal activists came to rely on their community group for social support, those least involved continued to rely on their families and relatives. Activists exceeded nonactivists in the loss of friends as well as in finding new friendships. That activists underwent the greatest change from the incident is understandable; they tended to be the most heavily impacted by the disaster (Stone and Levine, 1985).

Because activists are more absorbed in the issues, they become estranged from prior support networks and seek new "spatial" relationships with those sharing their ordeal. In contrast, those least affected have less reason to become activists, to disrupt existing relationships and, therefore, to seek new ones.

A Source of Information

Neighbors, as they interact and offer social support to one another, share information about the neighborhood (Unger and Wandersman, 1985). In Legler, such communication was minimal before the pollution crisis forced lines of intraneighborhood communication to be opened. Subsequently, the community had to contend with an abnormal informational context characterized by an extreme degree of uncertainty.[8]

The Concerned Citizens Committee helped to supply information to Legler residents at a time when they faced real confusion of facts and denial of responsibility for clarification on the part of responding government agencies. When respondents were questioned about sources of information during the water crisis, they gave government and the press negative ratings. Government at all levels—the township and higher—was not seen as open or truthful. The press was seen as depending upon the township government for information. Television was too sporadic in its coverage to be informative. Only the meetings of Concerned Citizens and its informal communications network were consistently seen by group members as being a reliable source of information. If anything suspicious happened in the neighborhood or if anything important was learned, this information was channeled immediately to the group's leadership. Perceptions were frequently verified with members of the executive board as rumors passed through Legler. The group's periodic meetings came to be seen as the one authoritative place where people could get straight answers to questions.

In a parallel vein, L. Gibbs (1982b) reported that the Love Canal Homeowners Association provided residents with information about chemical hazards, translation of scientific reports, and directions for filing relocation papers. Similarly, Baas (1986) described how community organizations served as the primary sources of information in four Dutch communities.

A Source of Power

The development of community organizations tends to provide people with a new sense of power in the midst of a situation that otherwise produces an overall sense of loss of control. There are at least two benefits from the community group's role as a path to power. The group serves as a collective means to achieve commonly shared individual goals. And it helps to reverse some of the psychological damage that occurs from the inherent powerlessness of the situation.

Community Power. As a source of collective power, the Concerned Citizens Committee in Legler helped people to overcome their sense of powerlessness in a difficult situation.

> We felt support. We were powerful enough not to have to take it. The township underestimated us "hicks." We made the township accountable.

By forming a group that could wield power and be effective, residents had a direction in which to channel their frustrations constructively.

> We helped each other not overreact. Some of us were going to township meetings with guns, we were that frustrated and angry. We felt like our arms were pinned back. We all channeled our frustrations into Concerned Citizens. Otherwise, there would have been vigilante violence.

Most importantly, the group allowed families to take actions that they might not have been able to undertake individually. Media coverage was sought and coordinated. The lawsuit was particularly mentioned by residents as a course of action made possible by their union. By turning to legal expertise, the Concerned Citizens Committee undertook to attain some balance of power between citizens and the experts brought forth by government and those responsible for the contamination. The lawyers, in turn, hired specialists able to address the issues faced by the Legler neighborhood.

In other instances, community groups have directly attracted experts to work with them in an advisory role. The experts helped them understand the significance of new developments, plot strategy, and even take on adversaries directly. The Love Canal Homeowners Association made extensive use of such expertise, some volunteer and some government-sponsored. Particularly important was the work of a cancer researcher, Beverly Paigen, in helping to reinterpret government data, develop the capacity to collect additional information, and interpret this information credibly inside and outside the neighborhood.[9] In the Netherlands, mobilization of expertise has been shown to be a key to community power, allowing citizens to demonstrate weaknesses in the reassurances given by the authorities and to put forward alternative plans of their own (van Eijndhoven and Nieuwdorp, 1986).

Community organizations are also able to create an infrastructure that supports needed activities. L. Gibbs (1982b) notes that the office of the Love Canal Homeowners Association served as a base for community participation. She stresses the importance for community organizations to provide such a setting.

Through their collective activities and ability to attract resources, community organizations provide a means for attaining the power needed to address the complex issues resulting from toxic exposure. Baas (1986) studied four Dutch communities where there was little confidence in the solutions proposed by government. In each case, community groups served their representative function well enough to earn overwhelming ratings of confidence from residents. These groups gained some control over community process by providing independent sources of information, criticizing technical investigations, and keeping authorities informed of local developments and the desired solutions.

Shaw and Milbrath (1983) note that for a community group to successfully influence policy, it must be able to put forward clear goals and to develop good strategy. The Concerned Citizens Committee in Legler was in fact successful in meeting its two major objectives; an alternative water system was installed and their lawsuit against Jackson Township was won. The Love Canal Homeowners Association succeeded in bringing political pressure to bear, resulting in a buyout of homes. Such success may not be uncommon. Thus, Freudenberg (1984a) found that 83 percent of the community groups he sampled perceived themselves to be at least somewhat successful. Half reported the elimination or reduction of the hazard that was the group's main focus. While these findings may have been subject to a sampling bias favoring successful groups, the importance of participation at the community level by toxic victims in affecting the course of events is clearly demonstrated.

Individual Power. Community organizations help residents to cope with what otherwise might be an unmanageable situation. As a result, the individual family is buffered somewhat from the ongoing stresses that result from dealing with government in seeking solutions to the crisis. A powerful grass roots organization also has benefits for the sense of security and control felt by its members. Rather than merely being dependent upon government to meet their needs, residents use the organization as a vehicle to regain some control over the response to the situation. Thus, while government might leave them in the lurch, Stone and Levine (1985, p. 158) note that residents' "own organizations would not consider the matter 'closed' as long as people were still having problems related to Love Canal." Psychologically, therefore, community organizations help to restore to victims a sense of control over their lives (Edelstein, 1982; Levine, 1982). Thus, "If we assume that people feel better about themselves after they meet difficult challenges with action, we can speculate that the organized citizens' groups provided a means for people to feel more in control of events, even though active membership entailed a change in personal behavior that was surprising to the actors themselves" (Stone and Levine, 1985, p. 173).

Toxic activism helps people recoup some lost psychological strengths. But it also produces personal growth, several examples of which were discussed in the previous chapter. Shaw and Milbrath (1983) suggest that leaders of Love Canal groups experienced enhanced self-control, self-worth, and personal efficacy. Stone and Levine (1985) similarly suggest an association between activism and positive individual change at Love Canal; nonactivists reported substantially more negative outcomes but did not appear to change their behavior during the incident. Activists learned to assert their rights.

There is, of course, a relationship between this growth and the ability of the organized groups to achieve their desired ends. Thus, power is heightened by an increase in the competence of the community group. While some expertise is achieved through the use of consultants, at least equally as important is the learning evidenced by community leaders (Stone and Levine, 1985). Therefore, the real importance of consultants to the citizens' groups may at times be less as technical experts than as educators. Individual learning translates to group power. The enhanced sophistication of citizens (Freudenberg, 1984a; Shaw and Milbrath, 1983) reflects this development of community organizations nationally.

The Consensus/Dissensus Continuum

There are also limits to the development of community as a response to toxic exposure. Community development is dependent upon the existence of an adequate level of "consensus," or agreement about the basic circumstances of exposure among residents of the contaminated area. The absence of consensus results in "dissensus," a situation with a high potential for conflict due to divergent definitions of the nature and origin of and response to the situation. Both natural and technological disasters are thought to be conducive to consensus (Quarantelli and Dynes, 1977). However, a more complex view is suggested by the toxic cases examined in this book.

First, dissensus marks the extracommunity boundaries around the contaminated area, where diverse institutional, corporate, and social forces combine to cause government inaction and the stigmatizing and blaming of the victims. The very nature of most contamination incidents challenges the economic foundations of society because industry is commonly the villain; Freudenberg (1984a) reports that the most common source of opposition to local organizations is industry. The overall result is a climate of conflict or dissensus.

Uncertainty and divergent pressures within the boundaries of contamination also result in the possibility of divisiveness among neighbors, reported in one-third of the communities studied by Freudenberg (1984a). At the same time, extracommunity conflict and the resulting isolation serve as impetus for agreement among the victims of exposure. As a result, among those sharing a common view of the severity of the toxic threat within the community, there is likely to be an emerging consensus that serves as the basis for organized community response.

Thus, consensus and dissensus both are likely to be found in exposed communities. Within a given community, it may further be useful to think of a continuum between dissensus and consensus rather than seeing these as dichotomous. As was seen in the Legler case, even incidents involving a fair degree of consensus are marred by conflict. Perhaps elements of both consensus and dissensus are likely to be found to varied degrees in different aspects of any toxic incident due to the inherent uncertainties of the situation.

The mix of consensus and dissensus may vary over the stages of toxic disaster. Community displaces family as the locus of decision making and action only at certain points in a contamination incident, such as when a community response is driven by the need for organized reactions to certain stages of government response (de Boer, 1986), particularly efforts to define solutions (ICF, 1981). If the motivating concerns are eventually addressed, the grass roots organization may well lose its raison d'être and its members return to more privatized concerns (de Boer, 1986). In other words, the group may literally put itself out of business by restoring the ability of private families to manage their own interests.

Furthermore, the lifescape changes discussed earlier run counter to community identification. While successful neighboring requires an attachment to place (Unger and Wandersman, 1985), feelings about home and environment are frequently inverted by the exposure experience. Thus, the new spatial network is undermined. Activism may take the form of a temporary community fighting for its own dissolution through government-sponsored relocation (as in the cases of Times Beach, Love Canal, and, more recently, Centralia). Even if the physical community is "repaired," further community development may be impeded. Thus, in Legler, a substantial proportion of the residents that I interviewed expressed a desire to move away if a successful lawsuit made it financially feasible to do so. Community development, in this case, may never broaden beyond the original focal issue, despite the spatial networking, simply because many residents have lost their long-term commitment to the neighborhood.

In this context, the community organizations that develop in response to such hazards as toxic exposure are unlike the permanent community organizations aimed at the maintenance of ongoing activities in a fairly stable environment. Instead, they resemble the kind of temporary organization described by Bennis and Slater (1968; see also Katz and Kahn, 1978) that forms only for the duration of its problem-solving effort.

However, the temporary nature of grass roots toxic organizations is affected by the fact that, in the preponderance of cases, toxic incidents move very slowly toward stages of solution. For example, unlike the Dutch incidents studied by de Boer (1986), there are relatively few places in the United States where government has implemented long-term remedial measures. As a result, the lifecycle of American community groups tends to be prolonged through a grueling period of study and evaluation. Groups become exhausted as they enter slow-moving and protracted stages of the incident during which continued mobilization becomes difficult to sustain. Lawsuits aimed at speeding the conclusion of the incident similarly tend to drag on beyond the coping resources of many victims' groups.

The point is that temporary organizations, created quickly in response to a toxic crisis, are most commonly forced to persist long beyond initial expectations. Such organizations are, in a sense, not voluntary. They are also grafted on top of the preexisting lives of the active members and thus are themselves a source of continuing disruption. It is not surprising, therefore, that elements of dissensus can be found in these organizations. The recognition of the lifecycle of community organizations can be useful from the standpoint of organizational management, helping to identify the points at which group members are most stressed.[10]

Finally, as reactive community groups succeed in responding to their original crisis, some are able to make the transition to becoming innovative and anticipatory organizations (see Botkin et al., 1979). Such organizations may even play a role in the discovery and recognition of additional instances of contamination, acting as community watchdogs in preventing further threats.

Consensus and Dissensus in Legler

The functioning of community groups responding to toxic exposure is subject to many influences, including the characteristics of the area, the residents, the political and geographic community, the specific toxic problem, the way the problem has been handled, and the characteristics of the members of the group (see Edelstein and Wandersman, 1987, for a full discussion). These variables help to determine the ability of the group to create and maintain an operating consensus. The Legler case study highlights some of the dynamics surrounding the consensus/ dissensus continuum.

Organizational Sources of Dissensus in Legler

Legler's Concerned Citizens Committee formed around a cluster of earlier residents who had previously fought the location of the landfill in their section. These leaders began to network almost immediately upon the announcement of the water contamination. They assumed a high profile and spoke at township meetings early in the incident. As issues crystalized and increased networking and attempts to share information occurred, the leadership expanded to include newer residents who, as a group, had been fairly isolated from long-term residents in the neighborhood.

Three of the seven key leaders, including the principal leader, were male. None of these men had demanding work schedules away from home (two were on disability and the third had the somewhat flexible schedule of a teacher). Therefore, from the standpoint of available time, they may have been more comparable to homemakers than to nine-to-five employees. In other respects, the key activists of Legler's Concerned Citizens Committee fit the profile of community leaders found at Love Canal (see Stone and Levine, 1985). They had young children, as did the majority of the group's constituents. The leaders also tended to reflect the group of residents who had come to Legler in the period from the mid-1960s through the early 1970s. Like their counterparts at Love Canal, they had sunk roots but were not old-timers.

In my interviews with the entire executive board and many members of the organization, it became clear that the brunt of the organization's work fell on a few key people (generally the executive board). Most of the other residents regularly attended general meetings but did little else. Therefore, it was not surprising that the leaders showed signs of extreme overload. They reported being blamed by the membership for emergent problems. Yet the larger membership expressed consistently positive regard for the leadership, citing their conscientiousness. That such support was not felt by executive board members suggests that somehow it was not being communicated upward within the organization.

This outcome can partly be traced to the organizational structure of the Concerned Citizens Committee. The major roles of the organization were invested in the executive board. Particularly after concerns developed that a spy was feeding information to Jackson Township about the citizen group's plans, the executive board became highly secretive. As board members became less open, their isolation contributed to their feelings of abandonment.

Also contributing to leader burnout was the course of the Legler disaster, which intimately involved the response of government. Initially, the group utilized broad participatory actions to capture attention and bring pressure to bear. But after the community entered the waiting period when decisions had been made but not yet implemented, the maintenance tasks required to keep the organization working toward its goals were handled best within the executive committee structure. Because they no longer were forced to go back to their membership for support, the leaders became increasingly isolated.

Conflict over Key Decisions

In Legler, the fact that an effective community organization emerged reflected an early consensus (at least among the approximately two-thirds of the total Legler families who joined the group) about certain common problems, goals, and approaches. But conflict was never absent from the picture. Conflict outside Legler's borders contributed to internal disagreement. Within the organization, conflict occurred at key decision-making junctures when differing goals among members surfaced. Besides these task-oriented issues that related to the organization's attempts to meet its goals, conflict was also evident in the socioemotional relationships within the group. That friction was a result of interpersonal tensions that developed over time. Two key decision points served particularly to bring conflict to the surface.

Wells Versus City Water. Legler was a community that lacked homogeneity in certain key ways. Most important was the division between earlier and later settlers. Those who arrived in Legler during the first major wave of development in the 1960s and early 1970s and those who had preceded them valued the rural ambience of the section and desired a high degree of seclusion and privacy from neighbors. They opposed further growth in the area. In contrast, most of the newcomers of the mid-to-late 1970s sought a more suburbanized "coffee klatch" community with opportunities for social contact among neighbors. They sought "development-style" living and the growth of amenities in the immediate area. Obviously, the potential existed for a basic clash of values.

As the Concerned Citizens Committee took shape around a shared need to respond to the water crisis, the organization was nearly split due to disagreement over whether to fight for a central municipally operated water district or for the digging of new, deeper individual wells. Generally, long-term residents favored the replacement of wells; newer residents thought that a "city water" system would provide the same degree of safety they presumed existed in their prior (usually urban) homes. To the former group, wells were a sign of independence and rural living; city water was a sign of government interference, additional expense, and chlorinated water. To the latter group, wells were a big responsibility and very uncertain (as the pollution experience confirmed); city water was a service that could (and should) be provided by government.

This conflict was resolved when the option of individual wells became unattainable. Local government claimed that public money could not be spent to improve private property. Furthermore, assistance was available to the township for building the more expensive central district project but was not available for digging individual wells (ICF, 1981). As a consequence, consensus was reestablished around the fight for city water.

To Sue or Not to Sue? A second major area of conflict developed around the initiation of the lawsuit against Jackson Township. This issue divided the neighborhood, eventually determining who belonged to the Concerned Citizens and who did not. Although my contact was only with residents who were in the lawsuit, it became clear from the interviews that diverse reasons accounted for the failure of about one-third of the families to join the lawsuit. Some refused to join because of their unwillingness or inability to risk an initial retainer fee for the lawyers. A few simply disliked lawyers and lawsuits. Others feared retaliation by the township either because they worked for government or because they feared that the certificates of occupancy issued for their homes might be revoked. Some felt that "you just can't win against city hall."

Others had personality conflicts with leaders of the community group; still others had difficulty facing up to the problem. Finally, some may have believed that they could share in the benefits of a lawsuit without having to take the risks. The question of participation in the lawsuit caused ongoing division in the community, although this conflict ceased to be internal to Concerned Citizens after nonparticipants in the lawsuit left the organization.

Strains from Prolonged Mobilization

It is not surprising to find that conflict occurred as the group entered its third year without having met its objectives. Leaders who had been mobilized for such an extended time began to show signs of fatigue. This burnout surfaced in the form of personality conflict rather than disagreement over issues. It is possible that this conflict was always present, but had been suppressed by the need for cooperation during crisis. But over time, feelings began to surface in arguments over the procedure to be used in meetings, the sharing of information among leaders and with members, and the extensive media exposure of some leaders.

The disagreements over wells and the lawsuit involved crucial task-oriented decisions. In contrast, the issues of trust and interpersonal conflict were socioemotional in nature, revealing the kinds of cumulative tensions that develop over time among a group of leaders under intense pressure. These conflicts surfaced at a time when Concerned Citizens had succeeded in its first objective of getting a new water source installed. There was less demand for members to pull together and to hide extraneous conflicts. The lawsuit, the major continuing concern of the group, was almost totally in the hands of the lawyers. With those external pressures reduced, the interpersonal relationships among the leaders began to reflect the suppressed issues that arise when a diverse group of possibly incompatible individuals are thrown together under high pressure. Thus, at a time when the leadership might have begun to relax, rising internal conflict within the group kept tensions at a boiling point.

These group conflicts may reflect a normal response to extended stress. Lang and Lang (1964) suggest that prolonged disaster strains the cohesive forces holding a group together and results in disruption due to collective emotional disturbances, displacement or mutual withdrawal, increased subgroup solidarity reflected in distrust of authority, charges of favoritism, and intensification of damages. Barton (1969) cites research to suggest that a "therapeutic community" that pulls people together to face a disaster will break apart when it must decide questions of allocation of aid or resources and when fellow sufferers discover that their association becomes less comforting and more painful over time because they continually remind one another of past crises. Additionally, in later stages of a toxic incident, personal rather than community issues become dominant, possibly contributing to a loss of the strong ties of mutual interdependence found earlier (de Boer, 1986).

Consensus and Dissensus Elsewhere

The degree of legitimacy attributed to the institutional context by the community group is another factor influencing the strain of prolonged organizational effort. In the United States, for example, government routinely downplays toxic risks and, therefore, its obligation to respond. As a result, community groups are forced into a confrontational mode with government, using the media and direct action to apply political pressure to override the regulators (e.g., L. Gibbs, 1982a and b; Levine, 1982). In contrast, community groups in the Netherlands enter into a semi-permanent negotiation structure with government that provides a legitimate forum for communication (Baas, 1986; van Eijndhoven and Nieuwdorp, 1986). While Dutch toxic victims still appear to distrust their government's response (Baas, 1986), the community group is accorded a legitimacy that rarely accompanies similar groups in the United States. In both cases, however, the internal development of the organization can be related to its acquiring a representative role with government (Baas, 1986; also ICF, 1981).

The Form of the Organization and Consensus

The form that a community organization takes can influence the development of consensus, as seen in an examination of three Dutch community organizations (van Eijndhoven and Nieuwdorp, 1986). One organization, in Griftpark, used a decentralized, consensual approach. The second, in Volgermeerpolder, delegated authority to experts. The third, in Merwedepolder, utilized a highly centralized and hierarchical structure. The first group relied on direct actions in which all members could participate, such as public demonstrations. The other two groups emphasized the use of petitions and public meetings. The limitations of the latter approach for maintaining consensus were most evident in Merwedepolder, where dissenting members carried out direct actions outside of the group, occupying a town hall and cutting bridge cables. The maintenance of consensus, as this comparison implies, requires continuing opportunities within the group for members to directly participate in activities that express their concerns, fears, and frustrations.

Several community groups formed at Love Canal in response to the toxic incident that began there in August 1978. The most prominent was the Love Canal Homeowners Association (LCHA). LCHA used media events to generate the political pressure necessary to force government action. Leaders attempted to speak for the community and were sometimes invited to government decision-making meetings. The group conducted its own epidemiological research with the expert help of a leading health researcher. Efforts to expand an initial government relocation effort to the outer areas of the neighborhood eventually succeeded through the organization's efforts (e.g., L. Gibbs, 1982a; Levine, 1982).

The LCHA bridged two of the Dutch models, being run principally by a core of key members and yet frequently undertaking broad participatory actions. Lois Gibbs, the organization's president and dominant force, deliberately used direct action as a means of controlling dissenters within her organization (L. Gibbs, 1982b). Fowlkes and Miller (1982) suggest that the strategies of the LCHA varied over time. The expressive direct action used early in the incident gave way to later activities that were more focused and instrumental (as evidenced by the attempt to halt remedial work at the canal through an injunction).[11]

The Dissensus Community—Centralia, Pennsylvania

Toxic disaster may also produce a dissensus so fundamental that interneighborhood conflict undercuts a long-standing sense of community. This is precisely the situation described by Couch and Kroll-Smith (1985; Kroll-Smith and Garula, 1985; Kroll-Smith and Couch, 1984) in their excellent study of Centralia, Pennsylvania. They profile a "community" (in the sense of municipality) in which there was neither an overall consensus nor the successful development of community groups representing different points of view. Thus, the benefits described for the development of therapeutic community were largely absent, and the level of stress was exacerbated by conflict. On the continuum, Centralia was truly on the dissensus side.

An anthracite coal deposit beneath Centralia ignited in 1962 when burning trash was disposed of in a strip-mined area. Subsequently, the uncontrollable fire continued to threaten the one thousand mainly elderly and ethnically diverse residents with toxic gases, subsidence, and explosion. With the crisis continuing into the 1980s, it is not surprising that factionalization occurred among residents. Kroll-Smith and Couch (1984) report disputes over whether the fire was really under the town and in what direction it was moving, over the best method to fight the fire, and over health and safety questions. Some residents were not threatened; others feared subsidence, explosion, and exposure to gases such as carbon monoxide, carbon dioxide, and methane. In this climate of uncertainty, government failed to clarify any of these issues. The fire was not located, a strategy for fighting it was not developed, and hazards were not defined. Existing confusion was exacerbated when standards for gas exposure were changed three times (Kroll-Smith and Couch, 1984).

Residents were angry neither at the fire nor at government. Instead, anger was directed inward within the community. Community violence was common. In contrast to the therapeutic communities found at Legler and Love Canal, "the primary stressor in Centralia is not the fire, but community conflict" (Kroll-Smith and Couch, 1984, p. 6).

Over a period of three years, seven different community groups developed, each representing factions in the general conflict over goals. The differences in the ways these groups tried to define the situation is well illustrated by these contrasting statements (Kroll-Smith and Couch, 1984, p. 7):

> Our message is clear. This town is in severe danger from an underground mine fire and each person in the town should acknowledge that danger and help us do something about it.

> We are being duped by the government and a handful of greedy people who want government to purchase their home so they can leave. There is no danger here (from the fire).

Internal conflict affected other communities without destroying the development of therapeutic community. Why then was Centralia so different? In Legler, internal conflict did not prevent the formation of one predominant community group to represent the bulk of the neighborhood. Similarly, at Love Canal, despite the presence of competing groups, the LCHA was the principal representative of the community. In neither neighborhood were all residents in agreement about the risks of exposure; however, community organizations were able to represent the shared interests of a substantial portion of the community.

Centralia differed in several ways from these other communities. First, boundaries were never firmly drawn delineating the endangered area. The drawing of boundaries in the other two cases served to define the range of danger, to activate anger at government for the way the boundaries were drawn, and to legitimate fears felt by residents living just outside the boundary who were at a disadvantage in receiving aid but who felt threatened by their arbitrary separation from the danger zone. Second, the danger in Centralia was even less actualized and more uncertain than in the other cases. Finally, the entire town was asked to agree to a strategy for addressing the concerns—a decision that properly belonged most to those proximate to the fire.

Kroll-Smith and Couch imply that divisions within Centralia corresponded to existing neighborhoods within the community. Not only were these neighborhoods the traditional basis of affiliation, but they experienced different degrees of exposure to the fire and thus different levels of concern. For residents of the neighborhood adjacent to the fire, support for a goal of government-aided evacuation was logical. However, this strategy threatened residents of neighborhoods less at risk who wanted to stay in Centralia. Thus, conflict was the result of trying to achieve consensus within a political community that was, in reality, several different communities of interest.

Given their insight, Kroll-Smith and Couch (1984) intervened to help create a basis for community discussion of the crisis. Neighborhood meetings were held rather than meetings of the entire community. Because people could meaningfully share issues at this level of participation, a process of discussion was begun that moderated the hostile atmosphere in the town.

As the Centralia example suggests, for community organization to be therapeutic there must be a shared sense of concern and a consensus around goals. Because "community" is such an elusive concept, dysfunctional attempts to create community support at a level that is beyond consensus may occur. Thus, in the absence of the isolation of the "victims," structures that represent both victims and nonvictims (both as self-defined and defined by others) can hardly be expected to serve as a basis for social support, information, and power on behalf of the victims. The failure of a unified community response in Centralia, therefore, offers support for the thesis that isolation of those at risk is a prerequisite for the kind of community organization found in most toxic incidents.

Isolation Without Community

Lying beyond the continuum from consensus to dissensus is another type of situation—contamination that affects only a few isolated families who must cope with it without collective local support. I have observed the consequences on several occasions. Pressures within isolated families are somewhat comparable to those within the families of leaders of larger community groups who become disproportionately involved in an issue to the detriment of their private lives. But the isolated family bears the full burden of work and is forced to develop the resources and play the roles that in other situations might be provided by many. They have no one with whom to share the constant pressure. And the sense that only fellow sufferers could understand their experience only underscores their isolation.

From the standpoint of managing a response to its situation, the isolated family suffers from a condition of "undermanning," signifying the presence of too few people to carry out the required tasks in a setting needed to achieve desired goals and maintain the system (Wicker, 1979). The isolated family must learn to play a wide range of roles, adapting to the demands of dealing with government and the need for developing new competencies. Struggling with their crisis, isolated victims are particularly subject to the strains entailed in activism.

The isolated family's ability to command power and attain information is substantially compromised by its undermanning. Regulators and politicians are reluctant to define a private woe as a community concern. Lawyers have little incentive to provide contingent-fee services. While media may see isolated victims as a "human interest" story, they are less likely to be "news." The importance of networking across incidents is particularly great for such families, who otherwise easily become convinced, in their isolation, that their experience of suffering is unique.

Concluding Comments on Enablement

While all communities face barriers to the creation and maintenance of grass roots organizations, the existence of a few communities ruled by dissensus does not disprove the basic model of community development proposed in this chapter. The extent of the development of community as a response to residential toxic exposure can perhaps best be seen by examining this phenomenon from a societal perspective.

Toxic Victims: A New Social Movement?

With thousands of communities responding to toxic threats across the United States, it is important to consider the cumulative effect of these grass roots efforts at a societal level of social process. What is the nature of the toxic victims movement, and what are its implications? Do toxic victims organizations represent an extension of the existing environmental movement, or do they form the basis for a new social movement?

The Emergence of a National Movement

Technological controversy commonly follows three stages leading to a mass movement. First, a warning about the technology is brought to the public's attention. Second, a small number of people at the local level oppose the technology (the way early Legler residents fought the siting of the Jackson landfill in their area). In the case of toxic incidents, most victims are drawn into controversy only after the discovery of exposure has been announced. Finally, a mass movement develops as "the growing number of communities experiencing disputes . . . serve as building blocks, all of a kind, which become linked together into a national coalition" (Mazur, 1981, p. 93).

A national toxic victims movement, consisting of exposure victims and those acting on anticipatory fear, has evolved precisely along these lines, building upon the networking of groups concerned with local issues (ICF, 1981). The movement has developed quickly in the decade since toxic incidents reached national prominence in the late 1970s. According to John O'Connor, director of the National Campaign Against Toxic Hazards, there is more activity in this country on the toxics issue than on any other single issue.[12]

The organization of this movement varies from that of the traditional social movement in which hierarchical bureaucracies are created at a national level (Blumer, as cited by Perry et al., 1976). Instead, the movement is highly decentralized, based upon active participation and networking, and, with few exceptions, has avoided the development of national leaders (Freudenberg, 1984b).

Although a decentralized structure puts opponents off guard, it also suffers from limitations. National corporations share a consensual view with other industries, pooling their knowledge and resources. In contrast, in a movement based upon local defensive battles against these corporations, different local organizations easily repeat each other's mistakes. And given their heterogeneous constituencies, community groups concerned with toxic exposure have a comparatively harder time developing a collective vision and strategy (Freudenberg, 1984b).

The movement that is evolving has begun to overcome the problems of decentralization while employing its strengths. Networking among community organizations sharing the same kinds of problems has occurred both spontaneously and as the result of deliberate organizing. While national-scale movements undergo an ebb and tide, affected by media coverage (Mazur, 1981; Molotch and Lester, 1975), the emerging toxic victims movement has a unique basis for being "refueled" due to the proliferation of new incidents that serve to broaden the grass roots base and to generate media coverage.

Furthermore, given their personal experience, members of the movement are unlikely to lose their commitment. For them, the issues are not global and abstract but personal and concrete. The movement builds upon these personal interests to force government action not forthcoming without the pressure of collective action (Morrison, 1983). The resulting media attention further has served to make environmental protection a top priority for nonvictims outside the movement (Milbrath, 1984).

The movement has developed the ability to supersede parochial concerns and move beyond local threats, such as the siting of specific landfills and resource recovery plants, to address the generic issues in waste disposal and treatment. As a result, the movement has begun to deal with what Lois Gibbs calls the "toxic merry-go-round" (CCHW, 1985). Rather than each community asserting "not in my backyard" or "take it somewhere else," many groups now demand "not in anyone's backyard" and argue for onsite solutions to toxic problems (CCHW, 1985). As a result, intersite conflict generated by the fact that waste hauled from one place is taken somewhere else has been minimized. Such conflict would hamper the growth of a national movement based upon consensus.

A number of factors have helped to spur the development of a national toxic movement. These include the development of facilitating organizations, supportive legislation, legal precedents, and continuing media coverage.

Facilitating Organizations. The creation of two national organizations focused on the toxic victim's plight has helped to give impetus to the movement.

The Citizens Clearinghouse for Hazardous Wastes, Inc. was founded in 1981 by Lois Marie Gibbs, former head of the Love Canal Homeowners Association. CCHW staffers have visited hundreds of groups to provide consultation and training and have worked in some capacity with some 2,500 grass roots organizations and two hundred traditional environmental groups. The organization's mailing list has grown to 10,000 names, with many of those listed holding membership. When group memberships are additionally considered, it is clear the organization represents an impressive number of people.[13]

CCHW's "Annual Report for 1985" (p. 1) emphasizes the goal of building a national movement.

> At Love Canal, we proved that when neighbors band together, do their homework, and stand up for their rights, they can win! When we won at Love Canal and then moved on to form CCHW, we started a movement that has since become the fastest growing and most dynamic movement that this country has seen since the Civil Rights movement. In fact, the fight against irresponsible hazardous waste disposal *is* a civil rights movement.

The organization disseminates information through a periodic newsletter entitled *Everyone's Backyard*, quarterly action bulletins, lots of face-to-face contact, and other publications. Covered topics include how to deal with government, experts, and lawyers; environmental testing; stress and burnout; standards for hazardous waste cleanup; and waste disposal methods such as incineration, deep-well injection, and landfills (CCHW, 1985).

Leadership programs for grass roots leaders are also offered. These programs require substantial commitment by local organizations. Three such programs were conducted in 1985, eight in 1986, and four in 1987. Additional programs involve organizing efforts in the South and for Hispanics, a technical assistance program that assists local groups in reviewing such documents as the plans for closure of Superfund sites, an organizing assistance program that provides advice to a dozen local leaders every day, and site visits around the country (CCHW, 1985). The organization has proven to be extremely innovative in addressing its concerns. Since 1985, for example, CCHW has held five roundtable conferences yearly to evaluate different issues. A landfill moratorium campaign has helped to discredit this waste disposal method. CCHW advocates the use of an environmental trust fund to protect neighbors of uninsured facilities, victimized by what CCHW calls the "NIMIC (Not in My Insurance Company) Syndrome" (CCHW, 1985). CCHW has also reached an international audience, spurred by its participation in a 1987 conference in Kenya.[14]

Although the action bulletins provide a means for applying political pressure, CCHW primarily serves as a support organization for smaller regional and local organizations. For example, when local leaders are in Washington, D.C., they lobby for their causes, with CCHW assistance, rather than having CCHW lobby on their behalf (CCHW, 1985).

In May 1986, CCHW held the "Fifth Anniversary of the Grass Roots Movement Against Hazardous Waste," attracting some four hundred community leaders to Washington, D.C., for twenty-six workshops on various topics. At the conference, ten resolution committees developed objectives for the toxic waste movement over the next five-year period. CCHW can thus be seen as actively building a national movement (CCHW, 1986).

A second national organization, The National Campaign Against Toxic Hazards, stems from the efforts of two other national groups concered with toxics issues, Citizens Action and the Clean Water Project. The campaign is waging a multiyear grass roots organizing effort on a state-by-state basis, building coalitions that can wield political muscle. While some activities overlap with CCHW's, the campaign is clearly more focused on the electoral and legislative process.[15] Its declaration of "citizen's rights" asserts many rights: the right to be safe from harmful exposure, the right to know, the right to cleanup, the right to participate, the right to compensation, the right to prevention, the right to protection, and the right to enforcement. These rights are to be secured through law (*Exposure*, 1984).

As a link in these developing national networks, statewide coalition groups have also been formed. While such groups vary widely in their form and activities, they represent an important step in building a grass roots movement. For example, the New Jersey Grass Roots Environmental Organization has become an active vehicle of support for local groups in that state. Originally a coalition of local groups, NJGREO has been able to achieve funding and to hire a full-time director.[16] Also in New Jersey, the Association of New Jersey Environmental Commissions publishes the *New Jersey Hazardous Waste News*. Vermont has been successfully networked by a group called VOC (Vermonters Organized for Cleanup), which has spawned local offshoot groups. In contrast, the New York Toxics in Your Community Coalition has faced a significant challenge in its effort to create an effective network in a much larger and more diverse state, but it has attained stability by linking to an existing statewide environmental organization.[17] Other statewide coalitions that have formed to address toxic exposure issues include the Arkansas Chemical Cleanup Alliance, the Louisiana Environmental Action Network, and the Maine and New Hampshire People's Alliances.

While the focus of the major national environmental organizations has been on litigating and lobbying, they have made some contribution to directly organizing and serving toxic victims (ICF, 1981). The Environmental Action Foundation published an excellent bimonthly paper, *Exposure*, for several years. Both Environmental Action and the Environmental Defense Fund have sponsored regional networking conferences across the U.S.[18] A number of organizations have published manuals for toxic activists, including the Sierra Club, the National Wildlife Federation, and the League of Women Voters.

Legislative Impetus. The original federal Superfund bill (The Comprehensive Environmental, Response, Compensation, and Liability Act of 1980, or CERCLA) which was expected to make possible the cleanup of the nation's most severe toxic sites, has been widely viewed as a major disappointment. Under the original law, insufficient money was provided for cleanup of sites. Instead, short-term remediation was offered. And even with this limitation, only a few sites received attention under Superfund. Lois Gibbs termed CERCLA a "deliberate distraction by government," inviting competition among communities for limited funding.[19] Adam Stern of the Environmental Defense Fund noted that hopes for Superfund to effectively address cleanup of contaminated sites gave way to frustration on the part of affected communities. Effective government action was blocked both by the inadequacy of technology and by political interference. This concern was greatly exacerbated by the record of the Reagan administration,[20] associated with instances such as the Stringfellow Acid Pits scandal during the Gorsuch administration at the U.S. Environmental Protection Agency (see Hill, 1984).

Against $1.62 billion in the first 5-year bill applied to work at 100 sites, the new Superfund bill passed late in 1986 targets 375 sites for the next 5-year period. This bill, known as SARA (Superfund Amendments and Reauthorization Act), will make $8.5 billion available for cleanups. Among its other improvements are provisions addressing victim compensation and the "right to know," an important tool for preventing toxic exposure. With strong environmental lobbying, it received wide congressional support but was nearly blocked by the Reagan administration's effort to protect industry from taxation (Crawford, 1986).

Ironically, the focus upon having a site listed for action under the federal Superfund may sap the initiative of state government to undertake cleanups. In New Jersey, for example, the State Spill Compensation Fund has not been used to fund cleanup of sites on the Superfund list. This failure to use state money for cleanups reflects more than a desire to avoid duplication, however. It indicates a desire not to exhaust available funding, since to do so would increase demands for corporate contributions.[21]

When local groups realize that inaction is all that they can expect from government, they are forced to form coalitions in order to lobby for help not forthcoming to isolated and nonpolitical local groups.[22]

Pressure from the Courts. The success of legal efforts to demand compensation and force remediation has also helped to spur the movement (see Morrison, 1983), although there have also been restrictive elements to key decisions.

Beyond a number of significant settlements, an early and influential court decision involved the suit by Legler residents against Jackson Township.[23] In 1983, a New Jersey jury found that Jackson Township had created a "nuisance" and "dangerous condition" in operating the landfill in a "palpably unreasonable" manner and that the landfill caused contamination of the plaintiffs' water. The verdict awarded to 339 plaintiffs more than $15 million for emotional distress ($2 million), medical surveillance ($8 million), diminished quality of life ($5.4 million), and reimbursement for costs such as hooking up to the new water system ($200 thousand). Awards granted to individual plaintiffs varied according to their proximity to the landfill, length and amount of exposure, and age.

A 1985 appellate decision upheld the award for loss of quality of life but reversed the awards for emotional distress and medical surveillance. In 1987 the New Jersey Supreme Court upheld the quality of life award, reinstated the medical surveillance award, and confirmed the reversal of the emotional distress award. The decision promises an important precedent, particularly because the reversal on emotional distress was due to a technicality in New Jersey law barring "pain and suffering" awards against government. The only limiting precedent involved the courts' refusal to grant damages for enhanced health risk, in part, because the risk could not be quantified and declared "reasonably probable" even if recognized as significant.[24]

A different successful use of the courts in New Jersey is evident in the practice of a former New Jersey Department of Environmental Protection lawyer, Michael Gordon. Rather than focusing upon compensation, Gordon specializes in seeking enforcement of environmental laws through the courts. In a string of successful toxic exposure and waste facility siting cases, he has acted on behalf of citizens and municipalities to get government action intended by state law.

Additionally, the National Environmental Policy Act (NEPA) demands attention to social and psychological impacts that can be shown to affect health. An important U.S. Supreme Court decision involving Three Mile Island (TMI) nuclear reactor I supports the need for consideration of such factors if a change in physical environment occurs due to a project. Although not a toxics tort, the case sets a precedent potentially providing citizens with a means of forcing their concerns to be addressed in the permit review of noxious or hazardous facilities (Jordan, 1984). At the same time, this case may involve a restrictive legal precedent. Psychological stresses due to the restart of TMI reactor I, which had been shut down during the accident at TMI-II, were found to have been caused by the fear of another accident, not by an actual accident at TMI-I itself. Thus, there was not a "proximate cause" for psychological stress associated with the start-up of TMI-I (Sorensen, et al., 1987). Justice Rehnquist's opinion sought to restrict the National Environmental Policy Act to examinations of physical impacts. Thus, in his words, "*risk* of an accident is not an effect on the physical environment" (*Metropolitan Edison and U.S. Nuclear Regulatory Commission v. People Against Nuclear Energy [PANE]*, 1983, p. 4373).

The outcome of such cases provides a mixed picture for the use of psychosocial concerns in seeking either compensation for toxic exposure or action through NEPA to address "anticipatory fears" from the siting of hazardous facilities. However, the courts and NEPA-type legislation have proven to be excellent forums for citizens to force waste companies and regulatory agencies to comply with laws.

Media Coverage. Extensive media coverage has brought awareness of toxic exposure into American homes. This awareness may account for the growing concern about toxic issues nationwide, captured by a variety of polls.[25] Such media coverage is often strategically valuable for community groups pressing certain demands. It also creates an "availability heuristic," whereby observers vicariously apply images associated in the media with one incident to others of a seemingly like nature (Slovic et al., 1980; also Molotch and Lester, 1975).

A New Social Movement or an Extended Environmental Movement?

Toxic exposure is a politicizing and radicalizing experience (Hamilton, 1985b; Molotch and Lester, 1975). At various levels, the safeguards and assumptions of a society fall aside. Victims are forced to develop a more critical perspective than is brought to most decisions by citizens. Yet toxic victims are not often "sectarian" environmentalists, acting out of a faith in ecological principles, but rather citizens reflecting the central values of society, filled with faith in the goals of a consumptive and capitalistic society (see Douglas and Wildavsky, 1982; also Milbrath, 1984). Toxic exposure gives them an unsolicited *de facto* critical environmental education. Their lives transformed by the incident, toxic victims are in a position to see what is nearly invisible for those for whom the system is working.

A New Constituency. The toxic victims movement is an environmental movement whose constituency varies greatly from that of the classic environmental movement (see Morrison and Dunlap, 1985; ICF, 1981). Geiser (1983) notes that unlike the professionals, scientists, students, and civic activists who led the environmental movement of the 1970s, the toxic movement is based in the middle-and working-class communities where toxic hazards can commonly be found. He further observes that

> This new movement is bringing forth environmental consciousness among people who were unlikely to think of themselves as "environmentalists." Because the movement is so tightly rooted in the immediate experience of people's community and family life, it has an urgency and a concreteness that is incredibly compelling. For these new "environmentalists," environment is not an abstract concept. It is something which has already exposed them to hazards which are debilitating them and hastening their deaths (p. 3).

Commenting in a similar vein, Newark community leader Bob Cartwright underscored the difference between the abstract issues of the first Earth Day and the new environmental movement based upon people who are "stuck with a problem and take matters into their own hands."[26] That the toxic movement involves a new constituency is seen both in the experience of staff of national environmental organizations[27] and in a survey of leaders of community organizations concerned with health issues (Freudenberg, 1984a).

The traditional environmental movement did not appeal to minorities. In contrast, Bullard (1984) suggests that toxic exposure of black citizens is motivating interest in the toxic victims movement. He details the specific threats to minority populations living "on the other side of the tracks." Examples can be found in rural areas of the South, such as Triana, Alabama, as well as in urban areas such as Dallas and Houston. In Bullard's analysis, these exposure situations evidence institutionalized racism in two senses. First, in Houston, a city that combines fast growth with no zoning, a *de facto* zoning pattern exists that sites waste facilities in minority neighborhoods. Second, Bullard suggests that all-black Triana, Alabama, the site of one of the worst toxic incidents in the country, has received relatively little media attention or government assistance compared with white communities such as Love Canal and Times Beach.

Bullard argues that the environmental movement is broadening beyond its white middle-class roots (see also Morrison and Dunlap, 1985; Cutter, 1981). The vulnerability of poor minorities to concentrated hazards has made pollution into a civil rights issue. At the same time, Bullard notes that blacks are often caught in a "jobs versus the environment" trade-off due to their economic dependence upon the manufacturing sector. He further argues that with a history lacking in politically oriented community organization, many black communities are subjected to a policy based on "the path of least resistance." In light of toxic victimization, however, Bullard looks to the development of black advocates for the environmental movement.

. Minority involvement in the toxic movement was evident at the CCHW national grass roots conference in 1986. For example, an Hispanic community leader was recognized for leading the fight that led to the third largest evacuation in the United States, the Ciudad Cristiana neighborhood in Humacao, Puerto Rico (CCHW, 1986). Strong minority leadership was also evident in two locations that I visited—Triana, Alabama, and Elizabeth, New Jersey.

Internationally, the toxic waste movement has expanded beyond the industrialized nations as Third World countries attempt to address the exportation of toxic products, pesticides, wastes,and manufacturing processes. In the aftermath of the Bhopal disaster, for example, environmental meetings and groups in India, Japan, Zimbabwe, Malaysia, Indonesia, Thailand, and Kenya commemorated the victims and planned actions to prevent further tragedies (Abraham, 1985).

"Traditional" Versus "Toxic" Environmentalists. Milbrath (1984) suggests that one-fifth of the U.S. population has abandoned the predominant American worldview to become "vanguards" for a "new environmental

paradigm" that espouses limits to growth, participation, risk minimization, appreciation of nature, and global consciousness. Another fifth adhere strongly to the old tenets, putting material wealth before a clean and safe environment. The remaining portion of the population are "environmental sympathizers," appreciating the need for change but not possessing an alternative to the dominant paradigm. They are sympathetic to the environment while still subscribing to material wealth and the other values of the "center."

Are toxic victims adherents of the new paradigm? Or are they simply individuals within the marketplace, raising the banner of "not in my backyard" merely to protect their private self-interest? Is there any environmental or ecological consciousness behind their positions? Unlike the more sectarian environmental public interest groups, their concerns are not global but specific and immediate. They seek personal protection, compensation for loss, and the ability to return speedily to the normative consumptive American lifestyle. They assert the rights of individuals to avoid toxic exposure. And they join together out of necessity, not voluntarily because of shared ideals (see Douglas and Wildavsky, 1982).

According to this analysis, one would expect to see major value differences between sectarian environmentalists and toxic victims newly radicalized but still cognitively part of the dominant growth paradigm. Lois Gibbs observes that there is indeed tension between what she terms the "yogurt" and the "bud" elements. While she sees both sharing the same ends, the traditional environmentalists and the grass roots community activists may vary in their methods. The latter know how to "street fight"; the former are given to compromise before the community activists would seek it.[28]

Gibbs reports instances of traditional environmentalists identifying with the same regulators distrusted by toxic victims. For example, a League of Womens Voters group told an inner-city organization that since their community was already so contaminated, it was the best site for new waste facilities. Traditional environmentalists also echo regulators in their attempts to avoid hysteria by being cautious in the release of information. Finally, environmental experts may distance themselves from community people by assuming knowledge of jargon used in the literature and in environmental legislation. Thus, after a lengthy lecture on "Superfund Contingency" presented to community people by a representative of a national organization, it became apparent that most of the audience had not understood the term but were reluctant to ask what it meant.[29]

Strategic goals and approaches may differ as well. According to Gibbs, environmental organizations attempt to work through the system: speaking to Congress, seeking to negotiate with business and government as equals, and making only "feasible" demands on government. When California toxic activists decided that they wanted expert assistance to be funded by Superfund, they were told by environmental groups that this was not feasible. They persisted anyway, winning their demands from the government.[30]

In contrast to these characteristics of conventional environmentalists, toxic victims use a grass roots approach based less upon lobbying than upon threatening unhelpful politicians with defeat. These new environmentalists are not afraid to cause hysteria and use it to garner power, seeing the sharing of information as an honest and effective approach despite the concerns raised. Rather than developing goals according to what can feasibly be attained, community groups reject compromise and go after what they believe they need and deserve. Community activists take pride in their refusal to accept waste projects, arguing that those proposing facilities should instead put them in their *own* backyards.

Several years ago, Lois Gibbs told a gathering of New Jersey toxic victims a story that dramatically illustrates the tension between toxic and traditional environmentalists. The scene was a Louisiana hazardous waste site hearing. Citizens set up an aquarium filled with contaminated drinking water from their wells. They loudly announced that the fish they were about to place in the tank would be dead by the end of the hearing. When the environmental officials and traditional environmentalists protested, the crowd began to chant "Kill the fish." Gibbs explained that this was not hysteria. "If we have to kill the fish to make the point, we'll do it. We're sacrificing our children." She later ended her talk noting that the traditional environmentalist would not have done what needed to be done—kill the fish.[31]

Asked whether this story indicated a rejection of an ecological perspective on the part of toxic victims, Gibbs indicated that it did not. She explained, "People are part of ecosystems too." In her view, a broader ecological perspective may follow toxic victimization, but only after people have been able to get their own needs met. Thus, she explained, when an issue such as the snail darter becomes the focal point of environmental concern, it illustrates to victims an unwillingness by environmentalists to come to grips with harder political issues suggested by dead babies in a toxic waste case. The former is viewed as a safe issue, not of major political significance; but the latter issue raises major questions of responsibility that are potentially expensive to address.[32]

Some indication of a broadening of issues is evident in the toxic victims movement. This is seen in the wider coalitions with other groups having related concerns (such as firemen and workers) and in the development of multi-issue advocacy groups that emerge as leaders of local single-issue groups mature (Geiser, 1983; also Freudenberg, 1984b). Similarly, Freudenberg found that some three-quarters of the responding groups have extended beyond their original concerns (Freudenberg, 1984a). However, it remains to be seen whether the long-term lifescape shifts found with toxic victims will bring them closer to sectarian environmentalists and, more importantly, will cause them to champion a more ecological perspective. Or will they remain a broad-based special interest movement?[33]

Freudenberg (1984b) suggests that the toxic victims movement needs a different vision. "In the decades to come the environmental movement will be battling for two fundamental rights: the right to live in an environment that does not damage health and the right to participate in making decisions about the environment in which one lives" (p.261).

Notes

1. An expanded version of the early part of this chapter, co-authored with Abraham Wandersman, appears elsewhere (Edelstein and Wandersman, 1987). The final section of the chapter has also appeared in another form (see Edelstein, 1984/1985).

2. See Erikson, 1976; Barton, 1969; and Wallace, 1957.

3. Edelstein, 1982; L. Gibbs, 1982b; Levine, 1982; and ICF, 1981.

4. Aspects of these community dynamics are discussed in Baas, 1986; de Boer, 1986; van Eijndhoven and Nieuwdorp, 1986; Stone and Levine, 1985; Freudenberg, 1984b; Creen, 1984; Shaw and Milbrath, 1983; Edelstein, 1982; L. Gibbs, 1982b; Levine, 1982; ICF, 1981.

5. See Baas, 1986; de Boer, 1986; van Eijndhoven and Nieuwdorp, 1986; L. Gibbs, 1985; Stone and Levine, 1985; Edelstein, 1982; and L. Gibbs, 1982b.

6. See M. Gibbs, 1986; Sowder, 1985; and Baum et al., 1983.

7. See Unger and Wandersman, 1985; see also Baas, 1986; and M. Gibbs, 1986.

8. See Fleming and Baum, 1985; Freudenberg, 1984a; Edelstein, 1982; Levine, 1982; and Slovic et al., 1980.

9. See Fowlkes and Miller, 1982; L. Gibbs, 1982a; and Levine, 1982.

10. A workshop on family stress by the Citizens Clearinghouse for Hazardous Wastes that I attended early in 1987 used the lifecycle of the community organization as a basis for projecting the type of stress toxic victims are likely to experience. Interactions of individual, family, and organizational dynamics were charted for different points in the evolution of a toxic incident in order to identify key issues for coping with each stage. Participants representing community groups from across the country then compared notes on the most effective coping strategies for organizational leaders, their families, and their groups.

11. In reviewing the manuscript, Levine took exception to the last point, suggesting that actions of the LCHA were instrumental from the onset.

12. John O'Connor, private conversation, May 21, 1984.

13. Lois Marie Gibbs, private conversation, May 15, 1984. Information was updated during a further conversation on December 10, 1987. See also CCHW, 1985. CCHW can be contacted at Post Office Box 926, Arlington, Virginia, 22216.

14. Lois Marie Gibbs, private conversation, December 10, 1987.

15. John O'Connor, private conversation, May 21, 1984. Note that the National Campaign Against Toxic Hazards can be contacted at 317 Pennsylvania Avenue, S. E., Washington D. C. 20003.

16. Madeline Hoffman, private conversation, May, 1986.

17. Ann Rabe, private conversation, May, 1986.

18. Adam Stern, private conversation, May 15, 1984.

19. Lois Marie Gibbs, private conversation, May 15, 1984.

20. Adam Stern, private conversation, May 15, 1984. See also Landy, 1986; Bowman, 1984; Davis, 1984; Novick, 1983.

21. Bob Cartwright, presentation to the New Jersey Grass Roots Environmental Organization, April 18, 1984.

22. John O'Connor, private conversation, May 21, 1984.

23. Steven Phillips, private conversation, December 15, 1986.

24. Steven Phillips, private conversations, December 15, 1986 and December 10, 1987. Also, *Ayers, et al.* v. *Township of Jackson* (A-83/84), decision of the New Jersey Supreme Court, May 7, 1987. See Judge Handler's partial dissent in the above matter for a critical discussion of the courts' distinction between "significant" and "reasonably probable" risk.

25. See Morrison and Dunlap, 1985; *Sunday Star Ledger* (Newark, New Jersey), March 11, 1984, and *The Times Herald Record* (Middletown, New York), February 6, 1984; see also Freudenberg, 1984a.

26. Bob Cartwright, presentation to the New Jersey Grass Roots Environmental Organization, April 18, 1984.

27. Adam Stern, private conversation, May 15, 1984.

28. Lois Marie Gibbs, private conversation, May 15, 1984.

29. Ibid.

30. Ibid.

31. Lois Marie Gibbs, presentation to the New Jersey Grass Roots Environmental Organization, April 18, 1984.

32. Lois Marie Gibbs, private conversation, May 15, 1984. Note that "deep ecologists" would be unlikely to accept Gibbs' humanistically biased definition of ecology (see, for example, Devall and Sessions, 1985).

33. Trent Schroyer, private conversation, May, 1984.

7

The Psychological Basis for a "Not-in-My-Backyard" Response

Introduction

This book so far has examined the impacts of past and present toxic exposure at various levels of social process. Now let us turn to the fear associated with the anticipation of future toxic exposure. Anticipatory fear is the basis for public opposition to stigmatized facilities, a phenomenon now known as the "Not-In-My-Backyard Syndrome" or NIMBY. NIMBY reactions have, at least on the surface, many of the same characteristics seen with toxic exposure—fears, stigma, stress, disablement, and community mobilization. People fear the kinds of lifestyle impediments that they have learned accompany toxic exposure. They anticipate the kinds of lifescape impacts that might accompany future toxic exposure.

Indeed, NIMBY is a legacy of the publicized experience of toxic victims. Over the past decade, knowledge of the consequences discussed in the prior chapters has made communities wary of potentially hazardous facilities. In case after case, there has been community resistance to the siting of stigmatized facilities, including landfills, hazardous waste disposal sites of all types, resource recovery plants, various sludge and septage facilities, nuclear waste disposal sites (as well as nuclear power plants), potentially hazardous industries, microwave towers, and power lines.[1]

Furthermore, one of the clear messages conveyed by past toxic disasters such as Legler and Love Canal is the abject failure of prior state-of-the-art disposal practices. In fact, some 80 to 90 percent of our hazardous waste disposal sites are recognized as unsafe; for every technology for waste disposal available, an array of environmental and human health hazards has been identified (Anderson and Greenberg, 1982). This inability to "solve" the "problem" of waste disposal has led to a general disbelief in the ability of engineers to create and maintain safe facilities. There are no more illusions of "secure" disposal sites.

Given this track record of prior disposal techniques, government agencies and entrepreneurs have found it increasingly difficult to site hazardous facilities (Anderson and Greenberg, 1982). Ironically, however, the failure to provide for adequate waste disposal contributes to the likelihood of toxic exposure. Tens of millions of tons of toxic substances enter the environment as unwanted waste every year (Purcell, 1982). This has led to one of the most-used clichés in environmental management—"The waste must go somewhere!" As the problem of waste disposal has acquired a priority position on municipal agendas across the U.S., the issue of public opposition has become the central concern (Weller, 1984). Some state and federal waste-siting procedures make it possible to override local opposition, but this is a question with significant moral overtones (Timmerman, 1984 a and b).

While waste disposal itself is a relatively small contributor to total toxic exposure, the crisis that has resulted from opposition to waste facilities is reflective of the general perception of toxic hazards. Thus, NIMBY is an important legacy of the social and psychological impacts of toxic exposure. Perception of the hazard is the key issue in instances of NIMBY. Thus, NIMBY is inherently a psychological phenomenon.

What does NIMBY mean? Macdonald (1984) notes that NIMBY may consist of conscientious concern, narrow selfishness, or a logical critique of an ill-conceived plan. NIMBY is not used as a value-free term. Thus, corporate and government officials refer to the second meaning when they speak of the "Nimby syndrome" as something of a social disease, a rabid and irrational rejection of sound technological progress. In contrast, community advocates portray NIMBY more sympathetically. Isaacs (1984) views NIMBY as a fiction invented to excuse poor planning. Because NIMBY implies that people are uninformed, Cornwall (1984, p. 9) prefers the term "LULU" ("Locally Undesirable Land Use"). Hayes (1984, p. 16) sees NIMBY as a fight for democratic rights more properly termed "NIMBI" ("Now I Must Become Involved").

In my view, the human reaction to perceived threat is not easily reduced to simple formulations. There are identifiable psychosocial bases for NIMBY. Furthermore, there is a rational basis for opposing potential hazards. Accordingly, any NIMBY incident is likely to involve elements of each of the definitions. I described earlier how Legler residents had opposed the municipal landfill when it was originally proposed, anticipating a range of problems. This chapter presents several additional case studies showing opposition to different types of facilities. All of the studies describe attempts to site waste disposal facilities, and all derive from my own research and observation. These case studies form a foundation for discussing both the psychosocial and the rational bases for the NIMBY response.

Case Studies of Community Opposition

Case One: Merion Blue Grass Sod Farm[2]

Early in 1979, local farmers became aware of heavy truck traffic into a sod farm at the edge of the fertile "black dirt" area of Orange County, New York. The black dirt is a major growing area for onions and other vegetables, as well as for lawn sod. Neighboring farmers began to detect foul odors from the site, which impaired their ability to work their fields and enjoy their homes. When town officials questioned the New York State Department of Environmental Conservation (DEC), they learned that the DEC had granted a temporary operating permit to a private firm, Nutrient Uptake, to spread sewage and septic wastes on the black dirt fields of Merion Blue Grass Sod Farm in order to fertilize the sod. The DEC had quietly determined that the application would have no adverse environmental impact. Because this "negative declaration" obviated the need for an environmental impact statement, the town had not been notified and citizens had not been given any opportunity to participate in the decision-making process.

When the temporary permit expired, the DEC arranged for an adjudicatory hearing to review the application for a long-term permit. Local communities and citizens could declare themselves to be parties to the hearing. Many did so. In fact, the town and its citizens were virtually unanimous in their opposition. Local farmers expressed many concerns about the sludge operation. They feared that word of the waste disposal would reach their customers, stigmatizing their crops. They were revolted at the substance being dumped, held the belief that it would contaminate food crops, and were personally concerned with the effects of contamination spread by soil, air, groundwater, and surface water.

The hearings ran throughout the summer and early fall of 1979. I attended most of the sessions, watching angry, frightened farmers fight for control over their community through a regulatory process that effectively disabled them. Among my observations were the following:

- An impressive expert on sludge flew in from Boston for one day to testify about how well the project was designed. Pointing to what he supposed was to be a sludge lagoon, he prompted the man next to me to jump up screaming, "That's my house! They're going to dump sludge on my house!" Then the expert's testimony about the attributes of various grasses was contradicted by sod farmers in the audience. The ability of the facility's berms to isolate wastes from the flood-prone Wallkill River was also challenged by residents well aware of the flooding characteristics. When I asked the expert about the hostile challenges from the audience, he retorted, "How dare these uneducated farmers interfere in an issue about which they lack competence! I'm the expert."

- The hearing was designed to minimize participation by residents. The sessions were held in daytime during the prime summer growing season, making it a real sacrifice for farmers to attend. The hearing officer and stenographer at the front were faced by tables at which lawyers and experts from the sod farm and several municipalities were seated. The public was seated to the rear of the hall. Lawyers controlled the proceedings with a wide array of objections and other legal maneuvers; the hearing judge had to continually remind them that this was not a courtroom. The frustrated public, forced to hold their comments and questions until after lengthy presentations and rebuttals by the experts and lawyers, displayed their anger at times, only to be told by the judge that respect was due because this was a courtroom.
- There was little opportunity for citizens to be heard and little credibility given when they were. While testimony from technical experts was carefully weighed in the DEC's decision to approve the facility, the local expertise of the farmers was essentially ignored.[3]

The hearing proved to be an inadequate means to involve the public. There was little attempt to assure that either the experts' arguments or the hearing procedures were clear and understandable. There was little to engender the residents' trust. Evident in statements by the judge, the lawyers, and the experts was the assumption that what technical experts said was "truth," while citizens held only private "opinions." Partially as a result, the hearing, like the facility, was never established as appropriate or legitimate for the community.

After considering the hearing record, the DEC granted the applicant a full permit to construct and operate the proposed sludge disposal facility. Farmers initially resisted, at one point staging an unsuccessful blockade of the entrance to the facility with their tractors. Their inability to stop the facility left the farmers despondent; a crucial community leader died of cancer depressed that she had been unable to stop the dumping.

But the farmers continued to monitor the site, helping to accumulate the evidence to close it. They also exposed as a "no-show" a DEC-appointed monitor who, assigned to inspect the site continuously, spent his working hours at a nearby striptease club but submitted time sheets for full workdays to the DEC. Ongoing questions regarding the dumping of illegal wastes, practices that diverged from the proposal, and permit violations plagued the facility for some six years before the DEC finally forced the operation to close.

Case Two: Al Turi Landfill, Inc.[4]

Al Turi Landfill, Inc., is located in Goshen, New York. The landfill is a private operation whose owners, according to government reports, are linked to organized crime (Hinchey, 1986). In the summer of 1980, hearings were held before an administrative judge from the New York DEC to review a draft environmental impact statement assessing the consequences of a proposed expansion of the facility. I undertook a study of the social impacts of the expansion on behalf of the town of Goshen, which sought to stop the project. The study involved group interviews with neighboring families selected by sampling residents within concentric rings drawn outward from the landfill site. Because Al Turi Landfill had been in existence on the adjacent property for more than a decade, I assessed existing impacts from that facility as a method for projecting future effects of the proposed expansion. Another massive landfill faced the proposed site along a major state road, suggesting the likelihood of cumulative impacts.

While farming was the predominant surrounding land use, some 1,000 people either lived or worked within an 8,000-foot radius around the proposed landfill.[5] Farm families had settled in the area because of its rich soil. Some of these families had roots dating back two hundred years. A number of professionals also lived in the area, often commuting to jobs an hour or more away. Most of these families had escaped the highly developed areas much closer to New York City for the privacy and isolation of a country home. Other area residents, clustered in a small development, had made residential choices more dependent upon quality of schools and proximity of jobs. All of the residents evidenced low mobility.

In considering the impacts of the operating landfill, these residents brought a number of different perspectives to the question of threat. Farmers were concerned about their herds and vegetables, not just their families. They were acutely aware that anything that affected the large amounts of water they drew from their wells could affect their dairies. In contrast, the nonfarm residents focused upon family issues and particularly upon their children's safety and health. Trucks and smells and dirt did not enhance their privacy and isolation. The landfill and related traffic had created fears that their children might be harmed.

The threat of water contamination undermined their hope that they had found a clean environment in which to raise children protected from environmental hazards.

As I interviewed residents living at various distances from the facility, I discovered that certain fears about the long-term consequences of the landfills were not based particularly upon proximity. The closest neighbors had documented concrete, daily impacts of the operation of the existing landfill that caused substantial erosion of their lifestyles. But even residents living as far as a mile from the site, who could neither see, smell, nor hear it, experienced fear. The fear was based principally upon the expectation that the landfills would pollute both the adjacent Wallkill River and the Southern Wallkill Valley Aquifer.

People were generally aware that the DEC had discovered a "contravention of groundwater standards" at the old landfill. Residents' concerns about expansion involved their inability to be certain of pollution moving underground and a belief that contamination had occurred and would occur. And while all of the best engineering drawings in the world plus promises of monitoring might encourage some level of confidence in the safety of the landfill expansion, there were no real guarantees. In fact, discussions of leachate from the new "high tech" landfill exacerbated concerns over what was leaking from the existing "old-fashioned" landfill. If pollution occurred, residents feared their health would be threatened and that they might be forced to move. That was happening to Love Canal residents at that very time as most of the residents knew from media coverage.

As a result of these concerns, residents gathered to plot strategy for opposing the proposed landfill during siting hearings. They worked closely with the town of Goshen in these efforts. Although they failed to stop the landfill, the effort led to the creation of an ongoing citizens group in the community, the Goshen Area Resources Association. This group has since expanded to other issues, but continues to monitor the safety of Al Turi Landfill, Inc. as one of its primary tasks.

Case Three: High Level
Nuclear Waste at the Richton Dome[6]

Under the Nuclear Waste Policy Act of 1982, the U.S. Department of Energy (DOE) commenced a search for at least one high-level nuclear waste repository. The Richton Salt Dome in Mississippi was one of the sites under consideration until 1986, when President Reagan designated three other sites for final consideration for the nation's first repository. Salt domes are an extremely stable geological environment, thought by some experts to provide ideal settings for nuclear waste during the tens of thousands of years that the wastes will have to be isolated in order to prevent the escape of harmful radiation.

The Richton Dome is in a rural but populous area some 50 miles upriver from the Gulf Coast. The project threatened to dislocate part of the small and close-knit community of Richton. The prospect of relocation was particularly frightening to elderly residents dependent upon their neighbors for assistance and their church for security.

Furthermore, as the result of publicity about a nuclear accident at the Tatum Dome in the 1960s, there was fear about radioactive contamination and distrust of government's ability to safely handle hazardous facilities. Opponents of the proposed Richton facility claimed that the entire population of the area would be at risk in the event of an accident at the site. Concerns about future health were associated with the project. There was additional concern that radioactive materials would leach into waterways and threaten the important coastal fishing industry. Their worst fears were portrayed in one newspaper article that pictured the repository destroyed by an earthquake in the next century, with fish along the Pascagoula River dying of radiation burns and residents of southern Mississippi being ordered to evacuate.

Among the projected impacts were traffic hazards along all road connections to the site leading from nuclear facilities across the United States. Locally, nuclear haulers would have to compete with log trucks on narrow, winding country roads, making a transportation accident a real possibility. Other concerns dealt with fears of short-term boom-bust growth during construction of the site as well as long-term induced growth. Meanwhile, the slow pace of the decision-making process resulted in a period of uncertainty during which plans could not be made, either by individuals and families or by the community at large. Even with the project now abandoned, there remains a feeling of "perpetual jeopardy" in Richton resulting from the likelihood that so visible a site will attract some other hazardous waste proposal.

Statewide and local groups formed to organize opposition to the facility. Politicians were pressured to lead the attack. One congressman, holding a piece of radioactive rock, was said to have initially claimed, "You can hold it in your hand; it is completely safe." A short time later he allegedly declared, "This project will go forward over my dead body."

Residents were angry at the defensive manner of DOE officials. They also questioned their competence. One resident claimed that she had spent eight years studying DOE documents, years that she said she would rather have spent with her grandchildren, and that the DOE experts could not match her local expertise.

I read that trash. They say that the wind doesn't come our way. Well, I know which way the wind comes; I don't think that the DOE is God Almighty.

Arrivals in Richton were greeted by large signs opposing the DOE project. One Richton official noted that this highly visible opposition stigmatized the town and scared away potential residents. Similarly, visible opposition in the coastal areas threatened to stigmatize the very fishing industry that residents feared would be contaminated.

The Richton project was viewed locally as an intrusion from Washington intended to punish Mississippi for past sins. Residents saw their state as stigmatized, viewed by northerners as "the rectum of the United States." Thus, beyond the other impacts, the facility was seen as an intrusion from the North. The social impacts may have helped to convince the Department of Energy to drop Richton as a site under consideration for the repository.

Case Four: The Vernon
Radium-Contaminated Soil Site[7]

In 1981 radium-contaminated soil was discovered in several Essex County, New Jersey, communities. Uranium mill tailings from a defunct luminescent watch and dial plant had apparently been used as fill for housing lots. Houses built atop the soil evidenced harmful levels of ionizing radiation from the radium and from radon gas decay. The New Jersey governor issued an executive order aimed at expediting the cleanup of the neighborhood.[8] With Superfund assistance through the EPA, the state Department of Environmental Protection (DEP) undertook a pilot project to remove the soil from beneath a few homes in Montclair. In 1985 families from these homes were relocated to small apartments during the excavation project. Neighboring residents remained in their homes, where various protective measures were implemented by the DEP. Once removed from the ground, the contaminated soil was placed in thousands of metal drums that were then "temporarily" stacked outside the homes until a permanent disposal site could be found.

But locating a disposal site proved to be more difficult than the DEP had anticipated. The nearby city of West Orange reacted negatively to plans to store some of the barrels in that community. Some of the drums were successfully moved for storage to another municipality, Kearny, New Jersey, before public pressure stopped the transfer. This left another 5,000 barrels stacked in the yards of the Montclair homes. Operating nuclear waste disposal sites in Nevada and Washington rejected the soil, in part because it was not radioactive enough to warrant using precious room in these facilites. But their rejection stemmed also from the politics of nuclear waste disposal—Nevada and Washington were indicating their displeasure at being saddled with most of the nation's nuclear disposal problem. Under legal pressure to act on the governor's executive order, the DEP revised a siting study it had hastily commissioned, selecting a Vernon, New Jersey, quarry, in the far north of the state near the New York border, as a disposal site for the soil.

In the summer of 1986, Vernon-area politicians were informed by the DEP of the decision. The quarry would be taken (by eminent domain if necessary) and united with adjacent parkland and the Appalachian Trail. On-site soil would be blended with the radium-contaminated soil and used to reclaim the quarry. The resulting soil mix would contain radium and radon at levels in line with existing background radiation for the Vernon area. From the DEP's perspective, the result appeared to be perfect—a major waste disposal problem would be solved, a scarred quarry made beautiful, and no resulting hazard created.

Vernon politicians were shocked. They had no warning. They were given no voice. The decision had simply been made without their knowledge or participation. The DEP felt justified in proceeding in this way because of the extraordinary powers given it by the governor to solve this problem and because it viewed its project as totally benign. To DEP officials, the material was "just *dirt*." They were quickly to discover that the rural and suburban residents of northern New Jersey and southern New York did not agree.

Local politicians viewed the project as an intrusion from Trenton, a message that their area was only peripherally important and so should take wastes that no one else wanted. Additionally, they rejected the DEP's use of the governor's order as an excuse to waive a siting procedure that would guarantee them some say in the outcome. In response, a powerful political coalition formed around a strong consensus that the central government had violated the democratic tradition.

Once the radium-contaminated soil disposal proposal was made public, the news drew a strong response from citizens. Organizers had no trouble recruiting volunteers willing to stop the project. Within two weeks of their announcement, DEP representatives faced a hostile crowd of more than three thousand people at the local high school. The prospect of violence forced the regulators to arrange for an escape route from the high school auditorium. As officials presented a glossy slide show about the project, the audience became increasingly angry. When the displaced Montclair residents were cited as a reason for moving quickly with the project, the audience, as a unit, broke into a caustic "aaw." One DEP official repeatedly claimed of the soil mix, "It's just dirt"— each time driving the crowd to near frenzy. The mayor of Vernon and a group of citizen leaders had to intervene to prevent an eruption. And then occurred possibly the most impressive scene of the meeting. Following scores of ranting politicians, only one citizen rose to speak— but she spoke on behalf of the entire crowd. Already citizens were sufficiently unified to allow their views to be represented in this fashion. The resulting "organization" was termed the "No-Name Group." The Vernon mayor also convened a "yellow-ribbon panel" to review the case.

Citizens raised a series of concerns about the project, including fears that the radium might leach into a major aquifer and contaminate downstream areas in New York State. Both farmers and residents became concerned. The New York Department of Environmental Conservation agreed that the risk crossed the state border. Extensive organizing in New York nearly matched the level of concern revealed in New Jersey. Both communities began to think of themselves as part of "The Valley." Large fundraising events were held involving entertainers living in the area. Citizens, towns, counties, and New York State entered an array of court cases against the New Jersey DEP, tying the project in legal knots.

Perhaps most telling, hundreds of citizens on both sides of the border joined small affinity groups where they learned non-violent protest techniques. Citizen leaders claimed that, on short notice, a thousand people could be called to block trucks bringing the soil, which would constitute a major political defeat for the governor. Although in court the DEP won the right to temporarily store the barrels at the site pending further review, New Jersey backed down. Some six months after its initiation. the project was abandoned.

During the short-lived crisis, there were reports of people's reluctance to purchase nearby land. Anticipatory fear about the escape of radium was rampant, with "hell no we won't glow" and "keep the soil out of our valley" signs and bumper stickers appearing everywhere. The level of mobilization was itself a major impact. Resulting social networks were a major gain.

In June 1987, the DEP repeated its Vernon siting blunder almost identically when it tried to move the radium-contaminated soil south to Jackson Township, near Legler. The citizen response mirrored that found in Vernon, again blocking relocation of the barrels. In mid-1987, test barrels of the soil were shipped to Oak Ridge, Tennessee, to be mixed with highly radioactive material, thereby qualifying it for disposal at Hanford, Washington.

While this action paved the way for removing the excavated soil from four homes in Montclair, thousands of barrels will remain in Kearny. To further intensify the crisis, well over one hundred additional families in several Essex County communities were informed in early 1987 of dangerous radium contamination beneath their homes, an advisory insensitively thrown on their doorsteps. The enormous cost and questionable feasibility of soil excavation, transportation, and disposal for the soil beneath these homes has forced the DEP and a state radon advisory board to explore options other than excavation. But residents have come to believe that nothing less than excavation will protect them. The dialectic of double binds between citizen and regulator continues. Nobody wins.

A Psychosocial Basis for NIMBY

One major lesson was suggested to me by each of the four case studies. The Merion Blue Grass Sod Farm case taught me that the decision-making process is a key factor in disablement. The Al Turi Landfill, Inc. case taught me that anticipatory fear can be learned from experiences in other communities. The Richton, Mississippi, case taught me the depth of importance of stigma in a siting controversy. And the Vernon, New Jersey, case taught me that an enabling response to disabling conditions can result in a broad-based and powerful sense of community.

We can draw upon many of the concepts discussed earlier to delineate some of the psychosocial reasons for a NIMBY response. These key elements will be briefly reviewed.

Stigma. Facilities identified as "hazardous" are inherently stigmatizing and thus undesirable. They are likely to be seen as threatening a community directly, by virtue of physical hazards, and indirectly, by virtue of a "courtesy stigma" whereby the community also becomes stigmatized because of its direct association with the hazard (see Goffman, 1963). Thus, as seen in all four case studies, stigma is both an attribute of the hazardous project and a potential adverse impact of the project on the community.

Courtesy stigma is inherently tied to image, since image is what is harmed by stigma. Therefore, a logical inference is that the degree to which stigma is feared will relate to the extent to which the hazardous facility contradicts the community image projected. An area seeking to attract tourists because of its setting, an agricultural area known for its wholesome products, a family residential area—all are vulnerable to the devaluing of image. Accordingly, fear of stigma is likely to be a motivating force in the community's reaction to the proposed hazardous facility.

Furthermore, the very nature of siting controversies is likely to emphasize the courtesy stigma. As more people know more about the project, more harm is done. Thus, as the controversy becomes highly publicized, the community acquires an increasing stake in successfully stopping the facility. Ironically, were the project developed quietly with little public awareness, courtesy stigma would probably be minimized (although stigma might be a possible outcome of later accidents or contamination).

Health Threat and Anticipatory Fear. Anticipatory fear is invited by concern about the future hazards associated with a facility. As discussed in earlier chapters, health concern, particularly for children's health, motivates a vigilant response that prompts the prospective victim to rehearse the worst outcomes possible from the facility. People vigilant about threats to their families and homes are likely to prefer mobilizing to stop a proposed facility before it is built rather than waiting to react to a dangerous condition brought about by the facility's operations. This is particularly true for women and for people generally expecting to exert control over hazards (Stallen and Tomas, 1985).

Inversion of Home and Territoriality. Threat to the sanctity of home as a psychological refuge and an economic investment results in "territoriality," or the defense of space. In the face of threat, the home is defended. While territoriality specifically reflects private concerns, those who have a sense of "ownership" over their community may oppose a facility out of a collective concern as well. In this case, opposition to a facility is a form of collective territorial defense.

Loss of Control and Disablement. The inability to prevent an unwanted change in the community challenges residents' sense of well-being. To suffer the impacts of the facility is to be reminded of their impotence. The result may be feelings of depression and a sense of being helpless and disabled. The way to maintain their sense of control involves activism aimed at preventing the facility from being sited. Such activism is particularly found in areas where residents are not newcomers, where they have a strong sense of community, where they think they have a chance of winning, and where they can fight collectively (Bachrach and Zautra, 1985).

Stress and Lifestyle Infringements. A facility may be expected to disrupt residents' everyday lives, causing stress both by direct physical stressors and by the realization that their lives are disrupted due to a stigmatized facility. Anticipating traffic, odors, pollution, litter, noise, and other impacts, residents have reason to oppose the facility (see also Cook, 1983).

Victimization and the Loss of Trust. In siting a facility, government is essentially selecting certain residents to be the victims of any adverse impacts caused by the operation. Even when the facility is privately operated, government is responsible for issuing permits. When the rights of nearby residents go unprotected, it is not surprising that distrust of government results. Government invites distrust in the way it handles siting cases. Furthermore, facility siting can be seen as an outside intrusion easily associated with past disputes.

Enablement and Vigilance. Cases of NIMBY tend to involve citizens who are vigilant—observing events, exploring alternatives, and evaluating decisions. As knowledge is accumulated, the citizens become competent to analyze the technical considerations involved in siting controversies. They also learn how to actively participate, often discovering that government officials are lax in their enforcement of regulations. As a result, vigilant citizens may become so enabled that they can drive the siting process.

The more that vigilant citizens learn about the hazardous project, the more they are likely to appreciate the undesirable characteristics. Therefore, unless there is some major benefit associated with the facility, they realize that they can only lose by its siting.

In a sense, opposition to stigmatized facilities represents a proactive response, as opposed to the reactive response of toxic victims after they have already suffered exposure. Residents are reacting to their anticipatory fears rather than to realized ones. In organizing to block the facility, they protect identity, health, home, lifestyle, and personal control and avoid victimization. Avenues of participation are created. NIMBY is thus an enabling process.

The Rational Basis for NIMBY

As the case studies reveal, for the prospective victim of a hazardous facility, a number of basic issues arise that are related to the perceived legitimacy of the siting process. These issues are central to the rationality of the NIMBY response. Concerned citizens are likely to ask three questions that influence their assessment of the legitimacy of the proposal: Have I been heard? How will I be affected? Why me? Each of these questions will be briefly examined.

Are My Concerns Being Heard and Addressed?

A major contribution to the anxiety of community members concerned about a proposed facility is the recognition that they cannot participate directly in a decision that may greatly affect them. Siting decisions are routinely made by government officials or by special siting boards charged with meeting the needs of the larger community. Fears raised at public hearings often are not weighed in balance with technical and political considerations by the decision makers. Lee (1984) cites Daneke's thesis that this devaluing of citizen input is due to the introduction of the public into a process originally designed for experts. An unidentified person cited in Armour (1984, p. 103) offers support for this view.

> When you are an environmentalist, I discovered, you become a lesser person because you care. When you have to stand before Boards and you're just a

citizen and not an expert, not much weight is put on your testimony. We felt that the more we got involved, the harder it became to be involved in a democratic process. We were allowed our democratic right to speak. But at the end of the first hearing, we didn't feel that it was fair and impartial.

This source further noted that a Canadian government letter about a facility hearing referred to the people of the town "twenty-three times as objectors, opponents and complainers." Clearly, this is how the citizens were viewed by government officials.

The resulting sense of public disenfranchisemet may contribute to a broad consensus against the proposed facility that cannot easily be compromised. Three contributing factors to this reaction deserve to be noted.

Qualitative Factors. The issues raised by opponents of hazardous facilities are hard to quantify. As a result, citizens often reject the legitimacy of quantification. Meanwhile, experts seek to find a calculus for reducing social concerns to numbers, if they consider them at all. The result is a gulf between pseudo-objective and pseudo-subjective arguments (Timmerman, 1984b; see also Mazur, 1981).

Similarly, it is often not recognized that arguments about the "goodness" of a principle are different from those over whether information is "correct." For example, the issue of how much risk of cancer someone should be subjected to is a fundamentally different issue from how to best gather data about a leaking hazardous waste site. While the separation of the moral and technical questions can be useful in clarifying NIMBY-type disputes, the schism reflects the failure of government technocrats to distinguish between purely emotional assertions and appeals based upon moral principles. Although qualitative, the later can be supported by reasons based upon general principles. Citizens frequently raise such moral issues rather than arguments of fact. The "goodness" of a particular facility fundamentally involves a different question than does the projected performance of its technical systems. There needs to be room for both types of questions to be raised (Timmerman, 1984a). This is a distinction that government technocrats have been slow to recognize.

Local Expertise. Citizens' perceptions of fact often vary from the "facts" as understood by the regulators and other experts. This, however, does not mean that citizen input is less correct than is expert testimony. Fischhoff and his colleagues (1982, p. 253) cite data suggesting that experts are not necessarily any more factual than citizens. They comment that

A case can be made that experts differ from lay people of comparable social origin and educational level, not in the way they think, but in the substantive knowledge they have at their disposal. It is not clear that this knowledge affords them any special advantage in going beyond the available data or into realms outside their expertise. Indeed, the price for acquiring such depth of field may be reduction in the breadth of their view. . . . The more perspectives involved, the more local wisdom is brought to bear on the problem.

A Varied Focus. Timmerman (1984b) distinguishes between two universes of thought on toxic waste disposal. Universe 1 involves the approach of a calculating bureaucracy, while Universe 2 involves a mutualistic community. Not only are these two universes likely to think about risk differently, but they also may be interested in different issues. Thus, NIMBY is not only a problem of perceived risk, but also of conceived risk. The citizen has a very different conception of risk than does the regulator, as discussed in Chapter 5. Some indication of this difference can be found in the literature on public understanding of air pollution. While citizens defined air pollution largely in terms of its effects (e.g., odor, haze and other visual characteristics, sore throats, breathing problems, and other health effects), experts looked to the chemical composition and technical causes of the pollution (Evans and Jacobs, 1981). This difference in definition of the problem makes it likely that the citizen and the regulator/expert will talk past each other.

Timmerman (1984b) argues that citizens use a "coherent rationality" in responding to the regulators' "instrumental rationality" (e.g., siting criteria). The citizens' rationality is based upon generally held tenets of acceptability. As a result, Timmerman (1984a, p. 38) notes that citizens can invoke two "trump rights": "(1) An absolute right not to have one's own health jeopardised for someone else's good without one's permission; (2) An absolute right not to be forced to put someone for whom one is responsible into jeopardy."

How Will This Project Affect Me?

Many of the direct and secondary impacts potentially associated with a hazardous facility can be anticipated; they serve as further grounds for rational opposition. These generally correspond to the impacts earlier noted for actual toxic exposure. Citizens assume that if an impact is possible, it *will* occur. Experts employ the opposite assumption. They expect that facility designs will work as planned and therefore problems are unlikely to develop. Yet citizen advocacy based on a worst-case analysis is vital because attempts to mitigate the effects will not be undertaken if they are not raised forcefully (see Mazur, 1981).

Three different studies offer a picture of the impacts people associate with hazardous facilities. Krawetz (1979) summarized concerns raised by the public in documented cases of hazardous-facility siting for Alberta Environment. In decreasing order of importance, these were human health effects, environmental effects, risk and safety issues, policy issues involving the need, justification and siting of a facility, site-planning factors, and site-specific quality of life issues. Anderson and Greenberg (1982) sampled local interest groups in a region of New Jersey on their rating of a range of potential physical and cultural impacts useful in siting a hazardous-waste disposal facility. The most emphasized siting criteria were water threat and proximity to a population center. Finally, in a comparative evaluation of perceptions of various types of hazardous facilities, Lindell, Earle, and Nealey (1981) found that there are commonly shared beliefs about the potential impacts of such facilities even among respondents varied in their occupational ties to hazardous industries and their involvement with environmental issues. Ratings of facility risk placed toxic chemical disposal facilities in the same high risk category as nuclear disposal and power facilities. These operations were associated by respondents with high levels of perceived threat, risks difficult to prevent, a threat of catastrophic accidents, the likelihood of causing many deaths over time, dreaded risks, overall risks that were seen as equal to or exceeding their benefits, and technological systems seen as in-sufficiently well known.

Why Me?

Equity in hazards exposure is a final issue likely to influence the citizen's reaction to stigmatized facilities (see also Kasperson, 1983). The fact that one is asked to bear the risks for others without sharing the benefits provides a sufficient basis for perceptions of inequity. Given this, it is unlikely that siting of a facility will be seen as fair and just.

The "Reverse Commons" Effect. A popular analogy for the pollution of our planet has been the "tragedy of the commons," whereby what all hold in common is destroyed by the private pursuit of self-interest (Hardin, 1968). But not only is the commons affected; the private property of others is often affected too. While we generally look to government to protect the public commons, the "owners" are most often responsible for protecting their private domain. NIMBY commonly involves the clash of this individual responsibility with government actions on behalf of the common good.

When facilities are sited for *concentrating* wastes (the term "disposal" is actually inaccurate), there is a different twist in the tragedy-of-the-commons theme. This "reverse commons effect" occurs when impacted parties are asked to bear disproportionate risks for society. The general good is served at their expense. Therefore, the willingness of individuals to sacrifice their own interests is called into play.

Who Bears the Risks? While siting a facility usually involves generating criteria for identifying the "best" sites (see Anderson and Greenberg, 1982), the most "feasible" sites are actually chosen. Thus, patterns in hazardous-facility siting are likely to reflect the general power dynamics of the society.

Several years ago, a representative of a major waste disposal company, Service Corporation of America (SCA), addressed an environmental conference that I attended. He explained that the optimal sites for hazardous disposal facilities were urban centers such as Newark, New Jersey (home of SCA's Earthline facility). Residents of such cities, he explained, benefited from working in chemical factories located there; thus, they should accept the risks of chemical waste disposal.

While there are many flaws to this logic,[9] it points to the kinds of rationalization used in facility siting to create an inequitable pattern of toxic exposure. Thus, Bullard (1984) identified "institutional racism" in the siting of hazardous facilities in black neighborhoods in urban areas of the South, such as Houston. The General Accounting Office, in a 1983 report, found the same pattern in eight southern states. For three of four hazardous facilities studied, the majority of local residents were black. Furthermore, at least one-quarter of the population in the four communities had an income below the poverty level, with the impoverished population being almost entirely black (GAO, 1983). Similarly, Greenberg, Anderson, and Rosenberger (1984) found that hazardous sites were located in areas where, compared with the dominant society, people were poor and marginal (younger, older, black, or foreign). They caution that siting of these facilities may have been due less to political reasons than to proximity to plants generating the wastes. Thus, rather than blatant discrimination, biases in siting may involve a subtle process of inequity (Bullard, 1984).

Saturation, Distribution, and Power. Seley and Wolpert (1983) suggest that equity issues arise from three general locational considerations in siting hazardous facilities. First, there are the actual impacts of the facility upon people and the community, including the duration of the impacts, the scale of the project, and the sensitivity of the location. They note that a community saturated with facilities may have less impact sensitivity to a proposed project than might an area having few facilities. Yet to take advantage of such adaptation is to create equity impacts.

There are also distributional effects that determine whether the facility makes the host community better or worse off in relation to the rest of society. Within the community, some may profit from the facility while the general costs are borne by the public. Thus, one must ask (Seley and Wolpert, 1983, p. 85): "(1) Are the costs borne by one group in order to spare larger groups (or society as a whole)? and (2), Is one group's equity more important than another's (that is, should we assume a compensatory or distributive role for equity)?" Finally, equitability is affected by power dynamics. Thus, as people become more vocal in opposing a certain facility, the siting effort is likely to shift toward a more powerless community.

It is the political nature of siting that inspired Timmerman (1984a, p. 2) to comment that "hazardous waste management has become one of the central stages (or 'battlegrounds') upon which a series of moral dilemmas (or 'battles') are being played out." In this light, he cites the philospher Dworkin's argument that while all may not be equal, all deserve equal and proper consideration in such a manner that people not be forced to accept conditions which force them to abandon their sense of equal worth (Timmerman, 1984b).

Responsibility for Wastes

The corollary of having a few bear the burden for the many is that most people can escape having to take responsibility for the ecological consequences of their wasteful practices. As wastes are increasingly hauled to more distant locations because of poor local planning and the failure to site local facilities, the problems addressed in this book are only moved elsewhere.

But why should communities take wastes from other areas? Why would people truly want someone else's wastes in their backyard? It is interesting that those who condemn NIMBY pay little attention to what might be called "YIYBY" (Yes, In *Your* Backyard), the tendency of people to want to foist their wastes off on others. In the absence of people having to come to grips with their own waste problems, there is little pressure for them to conserve, recycle, clean up, and reduce the waste flow. When wastes go somewhere else, there is no impetus to be responsible for the consequences of one's actions. This is a moral issue. Perhaps the fallacious concepts of "waste disposal" and YIYBY are really the problem, not NIMBY.

NIMBY, therefore, is potentially an important force for cultural change. While it may not have roots in ecological thought, its end point may well have ecological benefits. In isolation, NIMBY can be a tool for avoiding responsibility and forcing it elsewhere. As a social trend, NIMBY has a different implication. Waste reduction, recycling, source separation, the avoidance of unnecessary plastic packaging, and other conserving steps will occur, not because they are profitable, but because no one will want the resulting wastes. Accordingly, NIMBY is a force for radically altering the paradigm of environmental degradation.

Conclusion

NIMBY is a complex response. It represents reaction to the inherent stigma and induced anticipatory fear associated with projects. And it articulates citizens' frustrations over the manner by which projects are sited. It stems in part from the failure of regulators to take seriously the psychosocial impacts of facilities. Because these concerns are unlikely to be subject to mitigation and negotiation, the citizen rightly sees the siting question as an all-out, win-or-lose battle. In a rationalized siting process, citizens are robbed of their power when the discourse of technical experts is the basis for decisions. Citizens concerns are viewed as private and subjective. The result is that citizens must go outside the process, to the political sphere, if they are to "win." The difference between communities that can mobilize political influence and those that cannot is evident in where NIMBY is successful.

The solutions do not lie in further rationalizing the process, although firm scientific grounds for siting decisions are important. They lie instead in openness, communication, and empowerment. Psychosocial concerns must be addressed directly, with the same conviction now used for environmental impacts. Timmerman (1984b) suggests, for example, that if the criteria that have been assumed to be met during siting are later found to be violated, the facility should be forced to close down as an automatic condition of its permit. Such guarantees, along with a corollary opportunity for local citizens and communities to police the permit (for example, using experts paid for by the facility but controlled by the community), would help to give local residents a sense of control over hazardous facilities.

One wishes for a vigilant citizenry, but not one that blindly opposes all projects entailing a risk. However, it will take government's recognition of the legitimate basis for the NIMBY response before siting of hazardous facilities can be meaningfully addressed.

Notes

1. Similar resistance is also found regarding socially stigmatized facilities, including low-income housing, trailer parks, and community residences for the handicapped, incarcerated, substance abusers, ill (especially AIDS victims), emotionally ill, or aged (see Edelstein, 1987).

2. This case is based on my field notes and a review of the hearing record and commissioner's decision from the permit hearings for a sludge spreading facility at Merion Blue Grass Sod Farms held in Wawayanda, New York in 1979. The case is described more fully in Edelstein (1986/1987).

3. These conclusions were drawn from a content analysis of the hearing record and the commissioner's decision.

4. This case is based upon Edelstein (1980), a report prepared for hearings before the New York State Department of Environmental Conservation on behalf of Goshen, New York. I should note that in addition to my work on the social impacts of the facility, I represented the Goshen Environmental Review Board at the hearings and in meetings with concerned community members. I also became the first president of the community group that resulted from opposition to the facility, formed after the hearings.

5. A large state hospital immediately adjacent to the landfill accounted for the bulk of the population.

6. This case is based upon Edelstein (1985), prepared for the State of Mississippi Department of Transportation.

7. This case is drawn from firsthand observation, review of DEP documents, and interviews with a number of key informants. I should note that a non-profit organization that I am president of became involved in opposing the Vernon site, raising money for the New York-based community group fighting the site, and involving New York State in challenging New Jersey.

8. Executive Order No. 56, Governor Thomas H. Kean, December 4, 1983.

9. Among the flaws: those living near factories may not work in them; even if they do, they may not be the major beneficiaries; density surrounding the facility will be high, putting more people at risk; workers and nearby residents already bear the risks associated with manufacture, and should not also bear additional risks from disposal.

8

The Societal Implications
of Contaminated Communities

In this brief concluding chapter, I hope to place the earlier impact analysis in some larger perspective.

Roving Wastes

Throughout much of 1987, the Western Hemisphere was given a lesson in the limits of the Western view of waste. A barge carrying 3,000 tons of Islip, Long Island, garbage plied the Gulf of Mexico looking for a nation, state, or country that would accept it for disposal. Nobody wanted it. For a while, there were reports that the waste would be disposed of in Al Turi Landfill, Inc., in my hometown of Goshen, New York. What struck me as odd was that the public outcry could be so loud against this particular pile of trash when two major landfills in town daily accepted roughly as much tonnage, half of it imported from out of county, including from towns near Islip. There was clearly something special about the Islip barge garbage. It had been stigmatized. It no longer was anonymous trash. In acquiring an identity, the garbage threatened, in turn, to stigmatize the place that took it in. Moored in New York city, awaiting the outcome of a court battle over whether it's load could safely be incinerated, the barge even became a tourist attraction. It was in this sense that incineration was a means of not only destroying the garbage, but obliterating its identity as well.

A possibly overlooked element of the story is that the waste was being exported to begin with. Not only did no one else want Islip's waste, but Islip did not want its own waste. Long Island has come to recognize that landfilling is a threat to the sole source aquifer beneath its sandy (and thus permeable) soil. Since this aquifer is the principal source of drinking water, it must be protected.

At the same time that Islip's homeless barge searched for a berth, residents of northern New Jersey faced a major garbage crisis brought about by the lack of local waste disposal options and their dependence upon out-of-state sites where citizens are becoming as dubious as they are about future toxic threats. People in northern New Jersey are not willing to accept the risks of disposal for waste materials that they generate. They have become extremely sensitive to ecological, health, social, and economic impacts that might accompany a disposal facility. They will pay enormous amounts for others to accept the risks for them. One New Jersey county is even shipping its waste by rail some three hundred miles to Johnstown, Pennsylvania.

We seek increasingly distant disposal sites so that we can continue to degrade materials freely to a state of spent utility. Currently, we seem to have targeted Third World countries as our next generation of dumping grounds, selling hazardous products in their markets, exporting banned pesticides for poor farmers to use, literally taking waste materials overseas for disposal, and relocating hazardous industries to locations where there is little environmental or worker regulation (Norris, 1982).

What is the point? It is ironic that fast-food wrappers discarded by teenagers in Islip, New York, should travel thousands of miles. The irony is not that the trip failed to identify a disposal site, but rather that needless wastes are generated to begin with and that we then think that distant others will take over responsibility for the consequences.

Clearly there is a societal message in all of this.

The Engineering Fallacy

Early in this century, Americans came to view science, and particularly engineering, as clearing the path to progress. In a period of disillusionment with politicians and their inability to cope with a world of increasing complexity, it was thought that engineers offered awesome insight into the world of hard facts. While the "technocrats," the political movement that sought to elevate engineering to formal leadership in American society, failed as a political party, the reverence for the engineer's ability to solve problems became wedded to the American way. To a world of revolution, expanding technology, and bewildering change came the steady hand able to design progress (Akin, 1977).

Along with this transformation came an unfortunate piece of baggage, what I call the "engineering fallacy." This fallacy involves the assumption that problems can be solved unto themselves, isolated from the complicating factors and uncertainties of an overly complex world. By narrowing the problem, it becomes controllable (on paper at least); one speaks of solutions, since the thing to do with problems is to solve them. But what of problems that defy solution? Too often we assume that they are solved, only to realize later that they have reemerged, demanding new solutions. In this way, the era of indiscriminate dumping gave way to landfilling, or discriminate dumping. When landfilling subsequently came to be recognized as an abject failure (Abelson, 1985; Montague, 1984), dumping gave way to incineration and deep-well injection. With each shift, society is convinced that it has "solved" the problems of hazardous waste.

In his book *Steps to an Ecology of Mind*, Bateson (1972) pointed toward an alternative method to that of engineering. Any learning occurs within a context that can be approached through metalearning. To gain access to metalearning, one must first learn how to learn. Yet learning to learn is decidedly not on the educational agenda of a society enraptured with the control of the engineer's hand. Rather than learning to deduce and isolate, the metalearner always seeks to recognize the context of any problem. Thus, the problem not only exists in its own right, it is a reflection of something more basic (see also Botkin et al., 1979).

The Societal Meaning of Pollution

It is hard to recognize the basic truths of a society (referred to earlier as paradigms). They are so obvious as to be invisible. Either one must learn to seek them out, or one must grasp the underlying meaning by revolution born of the inadequacies of the narrower understanding. For some reason, most of us can accept that, as individuals, we engage in denial and rationalization. Yet it is harder to grasp the fact that societies similarly have well-shielded blind spots. Once this is realized, we can search for these points of collective denial. The crisis suggested by widespread toxic exposure calls attention to some of our shared distortions in thought.

The Polluting Consensus

So what is the meaning of pollution not just *for* society but *about* society? This subject was broached by the prescient Norwegian playwright Henrik Ibsen in his 1882 play, *An Enemy of the People*. The protagonist, Dr. Thomas Stockman, physician at the baths in a resort city, discovers bacteriological pollution in the water. He realizes that the contamination, traced to upstream tanneries, may explain a pattern of illness among the patrons of the baths. When Stockman develops a plan for a new water system to abate the pollution, he receives the overwhelming support of homeowners and the press, both of whom see him as an expert acting in their best interests. Given that his brother is mayor, he naively assumes that he will be able to move ahead and address the problem.

But Stockman has not yet learned about political interference. He is confronted by the politicians (including his brother) and by the big investors in the baths, who want the issue kept quiet for fear of stigma. When the public learns of the cost of Stockman's solution, the people side with the politicians. No one is willing to pay for the repairs, the losses during the construction period, and the decrease in tourism due to stigma.

As a public official, Stockman is caught in a bind. Is his duty to those who control his job or to the protection of public health? Faced with overwhelming community pressure to conform to the general definition of "good," Stockman has a disturbing realization. The people collude with the establishment. They define their welfare in terms of economic well-being, not in terms of public health. This collusion involves a greater desire to bury the bad news than to bury the pollution. The problem will go away if knowledge of the pollution goes away! Stockman has done more than to make a recommendation about how to solve a problem. Because his critique inadvertently touches the very nerve of the culture, he is transformed from hero to villain. The source of pollution is no longer the tanneries that contaminated the water; it is now Stockman himself!

But Stockman can no longer see the problem at the level of analysis used by his fellow townspeople. For him, it is not a matter of a trade-off between costs and benefits, accepted because mitigating the pollution is too expensive. To Stockman, the issue is that the majority eschews wisdom. By opting for short-run profit while accepting risk to others, the town's residents reinforce the legitimacy of the order that created the pollution problem to begin with. A community consensus exists behind the status quo, and Stockman will not be allowed a veto, even if he is correct. Pollution is less of a threat to the existing order than are unemployment, high taxes, and reduced profits.

The Meta-Issue

Stockman's experience highlights the meta-issue—why have we come to accept contamination as a necessary cost of the good life? Said another way—why have we come to accept the polluting life as good? Given its intrinsic challenges to our core assumptions, it is no wonder that people deny the meaning of pollution. We no longer deny the existence of pollution; instead we adopt the engineering fallacy—that pollution simply needs to be "cleaned up." Landfills or other technological systems can be designed to *securely* contain hazards; pollution is merely a technological problem waiting to be solved. This is societal denial!

Rarely do we admit that pollution results from our wasteful society, from actions in which we collude. We balk at the costs of preventing pollution, while assuming that sufficient resources will exist to carry out the much more costly and difficult exercise of cleaning it up after the fact. Because it is harder to quantify the costs of toxic exposure than the costs of preventing it, we allow it to continue. This is a choice that we all collude in, much as Stockman's townspeople colluded to preserve the status quo.

How do we begin to understand a deeper social, psychological, and cultural conflict raised by the "crisis" of pollution? Given the nature of our society, this crisis is visible in a number of ways. The growing social movement of toxic victims reflects, on one hand, the total disruption of lifestyle and lifescape that accompanies toxic exposure. For many of these victims, the perception that they have been exposed has fundamentally shaken their faith in the "American dream." The dominant social paradigm—promising as it does a world of technological progress, economic growth, and personal satisfaction—is much harder to take for granted. It has been opened to challenge. The naive trust in others and in the system gives way to anger, resentment, and a sense of having been wronged.

In the absence of an alternative paradigm, toxic victims are left to deal with toxic exposure in ways that force them to continue participating in the system that caused the pollution. Toxic activists seek "cleanup" and other engineering solutions. They press for health testing as a way of gaining some control over the problem. And they demand compensation for victims as an economic accounting for losses. While all of these are necessary and vital, they serve to institutionalize and legitimate as a problem what might otherwise be viewed as a fundamental crisis and, thus, a challenge to our modern, industrial way of life.

While victims form a social movement to pressure society to pay the costs of solving the problem, nonvictims have a somewhat different framework. The contradiction appears when nonvictims speak. In environmental opinion polls most people in our society indicate a great awareness of pollution and a willingness to pay the costs of prevention and cleanup. This, however, reflects their attitudes and not their values. Their lives are so compartmentalized that they live a lifestyle that supports the pollution habit, without even seeing the contradiction. Pollution is seen as an abstract "issue"—a problem to be solved—not as a personal problem to address.

The NIMBY response is possibly an even more reliable indicator than public surveys of the fact that toxic exposure is viewed as a threat. Seeing how disruptive toxic exposure is to victims, many people have reasonably decided to oppose stigmatized facilities. Accordingly, when the threat of pollution becomes personalized, the prospective victims may be motivated to act to prevent the source. From their perspective, the engineering fallacy becomes highly visible. Yet these opponents themselves collude in the necessity for pollution in many ways. Fighting a specific threat does not change the compartmentalized way of going about one's life. Thus, even as they oppose threatening intrusions, many NIMBY adherents profit from polluting activities through investments in corporate enterprises whose identity they may not even know. They consume large quantities of products produced and packaged so as to maximize pollution. They treat their lawns with poisons in order to imitate Astroturf. Their ability to achieve the "American dream" is intimately wedded to the economic and technological system that produces pollution.

People are left with what amounts to a territorial instinct. Pollution will occur. Victims will suffer. The challenge is to avoid being one of those victims. Thus, the stigma and blaming of the victim discussed earlier come not just from a derision of the people who failed to avoid such threats, but also from a denial of our role in victimizing them.

The Challenge of Facility Siting

Facility siting has become an incredible challenge. The technocrats define siting as a problem. They suggest engineered solutions for resolving various threats, even though such solutions may be recognized as only temporary. They rationalize the solutions using environmental and economic criteria. But then they confront their major problem: the tendency of neighbors to "irrationalize" their decisions by considering the "worst case scenarios" and seeing the weaknesses in the plans. Since hazardous technologies by definition are not subject to mitigation for worst case scenarios, they cannot be "rationally" sited from the perspective of those who would suffer should problems arise. In siting, what is rational for the many is irrational for the few.

Siting thus becomes a modern ceremony for selecting victims for sacrifice. We have eschewed human sacrifice on the altar as uncivilized. But we have instituted a more modern form of slow death by cancer that, silently dreaded by all, is equally quietly accepted by most when it stems from either general background sources or from personally caused pollution. But when an identifiable source of externally caused danger becomes apparent, unless strong economic incentives force us to accept the threat, we are likely to express our opposition.

Again, our society embodies contradiction. Based upon the nuclear family as individual entity seeking to profit for its own benefit, our society rests upon a firm belief in private property. This property provides us the basis for a home as a means to separate ourselves from the competitive world. It is a material means of accumulating both wealth and status. And it is an expression of the degree of control we possess over the earth and other humans. Threats to property are, therefore, deeply psychological in nature. Herein lies a dilemma. On one hand, our beliefs in individualism, in the right to property and profit, and in the ethical acceptability of acting solely in one's own best interests create the basis for the acceptance of polluting activities. On the other hand, those beliefs form the basis for strong opposition to collective attempts to technocratically solve this "problem."[1]

Democracy is being challenged by government's attempts to "disable" citizen participants and to create bureuacratic avenues that can assure successful siting. Citizen activists are now branded as terrorists for their opposition to facilities. But ironically, in focusing so strongly on NIMBY, we miss the fact that polluters have developed a parallel response. "Not in My Smokestack" (NIMSS) is now the reaction by corporate polluters to regulatory activities aimed at overriding their individual interests. Through political action committees, industrial lobbying groups, the sense of shared self-interest among the elite, and, finally, the collusion of the consumer, this NIMSS movement has gone comparatively unrecognized by government. It is instructive that while NIMBY is derided, NIMSS is viewed as enlightened self-interest. The "problem" is again the victim, not the polluter! The majority solves the problem by creating yet another definable "minority." The result is a growing totalitarian tendency on the part of a corporate-backed government, aimed at pacifying resistance to facilities necessary for the overall "good" of society (i.e., the high-consumption, high-profit society).

Toxic Exposure and Societal Change

The ability of a toxic victims movement to counter this trend remains to be seen. As isolated groups of victims or facility opponents, they have little chance. Yet the very "backyard" nature of the crisis supports this isolation; people are not motivated to define a crisis as existing unless they are themselves threatened. Other people's problems are exactly that. "Backyarders" in this way collude in the social dynamic that victimizes them.

As the movement develops and networks are created and organized around national organizations like the Citizens Clearinghouse for Hazardous Wastes, there is an interesting potential for demanding a change in the society. The notion of "Not in Anybody's Backyard" suggests the same core issues that Thomas Stockman inadvertently raised. But a movement cannot be isolated by society in quite the way that Stockman was. Thus, the movement has the potential to force fundamental societal change, whereas Stockman was subject to ostracism.

There are more fundamental changes afoot as well. What will the nature of these changes be? I can only speak to what I see as the true alternative to the existing paradigm. In line with the earlier discussion of a new environmental or ecological paradigm, the fundamental changes that are needed require a redefinition of the human relationship to nature. The view that humans can dominate nature, so deeply embedded in our culture (White, 1967) and our economic system (Polanyi, 1944), must be replaced by the realization that we are part of nature. Rather than having a schizoid relationship to the earth (Slater, 1974), that leads us to pay little attention to the impacts that we are causing, we must come to attend carefully to the effects of our materialistic way of life. When we collectively view the land as an exploitable resource and a dumping ground, we are hardly innocent when our actions return to haunt us (see Devall and Sessions, 1985).

Toxic victimization is only the identifiable symbol of what is happening to all of us. We must learn to respond supportively to toxic victims, but not because they are so dramatically different than we are. Rather, we are they.

Notes

1. It should be noted that in attacking the values of capitalism, I do not mean to imply that communist countries are less polluting. In fact, these countries appear to engage in unchecked technocracy, resulting in pollution problems that at least rival those in the west. Both systems embody the commodification of nature (see Polanyi, 1944). The commonality of this outcome, however, should not deter a careful reflection on our own cultural origins for pollution.

Bibliography

Abelson, Philip M. "Waste Management." *Science*, 228(June 7, 1985), 1145.

Abraham, Martin. *The Lessons of Bhopal: A Community Action Resource Manual on Hazardous Technologies*. Penang, Malaysia: International Organization of Consumers Unions, September, 1985.

Akin, William E. *Technocracy and the American Dream*. Berkeley: University of California Press, 1977.

Altman, Irwin and Martin Chemers. *Culture and Environment*. Monterey, Ca.: Brooks/Cole Publishing Co., 1980.

Anderson, R. F. and Michael R. Greenberg. "Hazardous Waste Facility Siting: A Role for Planners." *APA Journal*, Spring(1982), 204–218.

Armour, Audrey (Ed.). *The Not-In-My-Backyard Syndrome*. Downsview, Ontario: York University, 1984.

Baas, Leo. "Impacts of Strategy and Participation of Volunteer-Organizations of Involved Inhabitants in Living-Quarters on Contaminated Soil" in Henk Becker and Alan Porter (Eds.), *Impact Assessment Today, Vol. II*. Utrecht, The Netherlands: Jan van Arkel, 1986, 835–842.

Bachrach, Kenneth and Alex Zautra. "Coping with a Community Stressor: The Threat of a Hazardous Waste Facility." *Journal of Health and Social Behavior*, 26 (1985), 127–141.

Barr, Mason, Jr. "Environmental Contamination of Human Breast Milk." *American Journal of Public Health*, 71, No. 2(1981), 124–6.

Barton, Alan. *Communities in Disaster*. Garden City, New York: Doubleday, 1969.

Bateson, Gregory. *Steps to an Ecology of Mind*. New York: Ballantine Books, 1972.

Baum, Andrew, Raymond Fleming and Jerome Singer. "Coping with Victimization by Technological Disaster." *Journal of Social Issues*, 39, No. 2(1983), 117–138.

Baum, Andrew, Jerome Singer and Carlene Baum. "Stress and the Environment." *Journal of Social Issues*, 37(1981), 4–35.

Beck, Kenneth and Arthur Frankel. "A Conceptualization of Threat Communications and Protective Health Behavior." *Social Psychology Quarterly*, 44, No. 3(1981), 204–217.

Becker, Franklin. *Housing Messages*. Stroudsburg, Pa.: Dowden, Hutchinson and Ross, Inc., 1977.

Bennis, Warren and Philip Slater. *The Temporary Society*. New York: Harper and Row, 1968.

Botkin, J. W., M. Elmandrja and M. Malitza. *No Limits to Learning: Bridging the Human Gap*. New York: Pergamon Press, 1979.

Bowman, Ann O'M. "Intergovernmental and Intersectoral Tensions in Environmental Policy Implementation: The Case of Hazardous Waste." *Policy Studies Review.* 4(1984), 230–244.

Brickman, Philip, Vita Carulli Rabinowitz, Jurgis Karuza, Jr., Dan Coates, Ellen Cohn and Louise Kidder. "Models of Helping and Coping." *American Psychologist,* 37, No. 4(1982), 368–384.

Brown, Michael. *Laying Waste: The Poisoning of America by Toxic Chemicals.* New York: Pantheon, 1980.

Bullard, Robert. "The Politics of Pollution: Implications for the Black Community." Paper presented at the Annual Meeting of the Association of Black Sociologists, San Antonio, Texas, August, 1984.

Burton, Ian, Robert Kates and Gilbert White. *The Environment as Hazard.* New York: Oxford University Press, 1978.

Campbell, Angus. *The Sense of Well-being in America.* New York: McGraw Hill, 1981.

Carson, Rachel. *Silent Spring.* Boston: Houghton Mifflin, 1962.

CCHW [Citizen's Clearinghouse for Hazardous Wastes, Inc.]. "The Five Year Plan of Action." *Everyone's Backyard,* 4, No. 3(1986), 5–8.

CCHW. "Annual Report for 1985." Unpublished report.

Cook, J. "Citizen Response in a Neighborhood under Threat." *American Journal of Community Psychology,* 11(1983), 459–471.

Cooper, Clare. "The House as a Symbol of Self." Working paper #120, Institute of Urban and Regional Development, University of California, Berkeley, May, 1971.

Cornwall, George. Comments cited in Audrey Armour (Ed.), *The Not-In-My-Backyard Syndrome.* Downsview, Ontario: York University, 1984, 8–10.

Couch, Stephen and J. Stephen Kroll-Smith. "The Chronic Technical Disaster: Toward a Social Scientific Perspective." *Social Science Quarterly,* 66(1985), 564–575.

Coyer, Brian Wilson and Don Schwerin. "Bureaucratic Regulation and Farmer Protest in the Michigan PBB Contamination Case." *Rural Sociology,* 46, No. 4(1981), 703–723.

Crawford, Mark. "Toxic Waste, Energy Bills Clear Congress." *Science,* 234(1986), 537–538.

Creen, Ted. "The Social and Psychological Impact of Nimby Disputes" in Audrey Armour (Ed.), *The Not-in-my- Backyard Syndrome* Downsview, Ontario: York University, 1984, 51–60.

Cutter, Susan. "Community Concern for Pollution: Social and Environmental Influences." *Environment and Behavior,* 13(1981), 105–124.

Dalton, Edward. "Combating Disease and Pollution in the City" in Donald Worster (Ed.), *American Environmentalism: The Formative Period, 1860-1915.* New York: John Wiley and Sons, 1973, 133–149.

Davis, Joseph. "Superfund Contaminated by Partisan Politics." *Congressional Quarterly,* 42(1984), 615–620.

de Boer, Joop. "Community Response to Soil Pollution: A Model of Parallel Processes" in Henk Becker and Alan Porter (Eds.), *Methods and Experiences of Impact Assessment*, special issue of *Impact Assessment Bulletin*, 4, No. 3/ 4(1986), 187–200.

deCharms, R. *Personal Causation*. New York: Academic Press, 1968.

Defoe, Daniel. *A Journal of the Plague Year*. New York: New American Library, 1960.

Devall, Bill and George Sessions. *Deep Ecology: Living as if Nature Mattered*. Salt Lake City: Peregrine Smith Books, 1985.

Dickens, Charles. *Hard Times: For These Times*. New York: Books Inc., n. d.

Dickson, David. "Limiting Democracy: Technocrats and the Liberal State." *Democracy*, 1, No. 1(1981), 61–79.

DiPerna, Paula. "Leukemia Strikes a Small Town." *The New York Times Magazine*, December 2, 1984.

Division of Epidemiology and Disease Control. "Groundwater Contamination and Possible Health Effects in Jackson Township, New Jersey." Report to the New Jersey State Department of Health, July, 1980.

Douglas, Mary and Aaron Wildavsky. *Risk and Culture*. Berkeley: University of California Press, 1982.

Drotman, D. P. "Contamination of the Food Chain by PCB from a Broken Transformer." *American Journal of Public Health*, 73(1983), 302–313.

Eckholm, Eric. *Down to Earth*. New York: W. W. Norton, 1982.

Edelstein, Michael R. "Towards a Theory of Environmental Stigma" in Joan Harvey and Don Henning (Eds.), *Public Environments*. Ottawa, Canada: Environmental Design Research Association, 1987.

_____ . "Disabling Communities: The Impact of Regulatory Proceedings." *Journal of Environmental Systems*, 16, No. 2(1986/1987), 87–110.

_____ . "Toxic Exposure and the Inversion of the Home." *Journal of Architectural and Planning Research*, 3(1986), 237–251.

_____ . "Psychosocial Impacts on the Community." Document prepared for the socioeconomic impact study of the proposed Richton, Mississippi, High-Level Nuclear Waste Repository, Mississippi Department of Transportation, 1985.

_____ . "Social Impacts and Social Change: Some Initial Thoughts on the Emergence of a Toxic Victims Movement." *Impact Assessment Bulletin*, 3, No. 3(1984-1985), 7–17.

_____ . "Stigmatizing Aspects of Toxic Pollution." Report to the law firm of Martin and Snyder, 1984.

_____ . "The Social and Psychological Impacts of Groundwater Contamination in the Legler Section of Jackson, New Jersey." Report to the law firm of Kreindler and Kreindler, 1982.

_____ . "Social Impacts of Al Turi Landfill, Inc." Report prepared for the town of Goshen, N.Y., July, 1980.

_____ . "The Role of the Physical Setting in Interpersonal Interaction." Unpublished manuscript, Fall, 1973.

Edelstein, Michael R. and Abraham Wandersman. "Community Dynamics in Coping with Toxic Exposure" in Irwin Altman and Abraham Wandersman (Eds.), *Neighborhood and Community Environments*. New York: Plenum Press, 1987.

Edelstein, Michael, Joel Kameron, Matina Colombotos and Syrrel Lehman. "Psychological Impact of Traffic and Attendant Factors of Air Pollution, Noise and Safety" in Richard Graham and Steven Posten (Eds.), *An Applied Natural Resource Inventory of the Borough of Paramus, New Jersey.* Paramus: Paramus Environmental Commission, 1975.

Erikson, Kai. *Everything in Its Path.* New York: Simon and Schuster, 1976.

Evans, Gary and Stephen Jacobs. "Air Pollution and Human Behavior." *Journal of Social Issues,* 37(1981), 95–125.

Exposure. Jan-Feb, 1984.

Finsterbusch, Kurt. "Typical Scenarios in Twenty Four Toxic Waste Contamination Episodes." Paper presented at the Annual Meeting of the International Association for Impact Assessment, Barbados, June, 1987.

Fischhoff, Baruch, Paul Slovic and Sarah Lichtenstein. "Lay Foibles and Expert Fables in Judgments about Risk." *The American Statistician,* 36, No. 3(1982), Part 2, 240–255.

Fitchen, Janet. "Cultural Factors Affecting Perception and Management of Environmental Risks: American Communities Facing Chemical Contamination of Their Groundwater." Paper presented to the Annual Meeting of the Society for Applied Anthropology, Washington, D.C., March, 1985.

Fleming, India and Andrew Baum. "The Role of Prevention in Technological Catastrophe" in Abraham Wandersman and Robert Hess (Eds.), *Beyond the Individual: Environmental Approaches and Prevention.* New York: Haworth, 1985, 139–152.

———. "Chronic Stress in Residents Living Near a Toxic Waste Site." Paper presented to the Eastern Psychological Association, Baltimore, Md., 1984.

Forester, J. "Critical Theory and Planning Practice." *APA Journal,* July, 1980, 275–286.

Fowlkes, Martha and Patricia Miller. "Love Canal: The Social Construction of Disaster." Report to the Federal Emergency Management Agency, 1982.

Francis, Rebecca S. "Attitudes toward Industrial Pollution, Strategies for Protecting the Environment, and Environmental-Economic Trade-offs." *Journal of Applied Social Psychology,* 13(1983), 310–327.

Freedman, Tracy. "Leftover Lives to Live." *Nation,* 232, No. 23(1981), 624–627.

Freudenberg, Nicholas. "Citizen Action for Environmental Health: Report on a Survey of Community Organizations." *American Journal of Public Health,* 74, No. 5(1984a), 444–448.

———. *Not in Our Backyards.* New York: Monthly Review Press, 1984b.

Fried, Marc. "Grieving for a Lost Home" in L. J. Dahl (Ed.), *The Urban Condition.* New York: Basic Books, 1963, 151–171.

GAO (U.S. General Accounting Office). "Siting of Hazardous Waste Landfills and Their Correlation with Racial and Economic Status of Surrounding Communities." Report issued June 1, 1983.

Geiser, Ken. "The Emergence of a National Anti-Toxic Chemical Movement." *Exposure,* No. 3(February 1983), 7.

Gibbs, Lois Marie. "The Impacts of Environmental Disasters on Communities." Reprint of the Citizens Clearinghouse for Hazardous Wastes, 1985.

_____. *Love Canal: My Story.* Albany: State University of New York Press, 1982a.

_____. "Community Response to an Emergency Situation: Psychological Destruction and the Love Canal." Paper presented at the American Psychological Association, Washington, D.C. August 24, 1982b.

Gibbs, Margaret. "Psychological Dysfunction as a Consequence of Exposure to Toxics" in A. Lebovitz, A. Baum, and J. Singer (Eds.), *Health Consequences of Exposure to Toxins.* Hillsdale, N.J.: Lawrence Erlbaum Associates, 1986, 47–70.

_____. "Psychological Dysfunction in the Legler Litigation Group." Report to the law firm of Kreindler and Kreindler, 1982.

Glass, D. and J. Singer. "Behavioral Aftereffects of Unpredictable and Uncontrollable Aversive Events." *American Scientist,* 60(1972), 457.

Goffman, Erving. *Relations in Public.* New York: Harper and Row, 1971.

_____. *Stigma: Notes on the Management of Spoiled Identities.* Englewood Cliffs, N.J.: Prentice-Hall, 1963.

Gottlieb, B. *Social Networks and Social Support in Community Mental Health.* Beverly Hills, Calif.: Sage Publications, 1981.

Greenberg, Michael. "Does New Jersey Cause Cancer?" *The Sciences,* (Jan.-Feb., 1986), 40–46.

Greenberg, Michael, Richard Anderson and Kirk Rosenberger. "Social and Economic Effects of Hazardous Waste Management Sites." *Hazardous Waste,* 1(1984), 387–396.

Greenberg, Michael, Frank McKay and Paul White. "A Time-series Comparison of Cancer Mortality Rates in the New Jersey-New York-Philadelphia Region and the Remainder of the United States, 1950–1969." *American Journal of Epidemiology,* 3, No. 2(1980), 166–174.

Gruber, Sheila. "Terrible Impact: Residents Suffer Psychologically from Living Near Toxic Site, Study Reveals." *The Ann Arbor News* (Michigan), February 10, 1985, A-17.

Habermas, Jorgen. *Communication and the Evolution of Society.* Boston: Beacon Press, 1979.

_____. *Toward a Rational Society.* Boston: Beacon Press, 1970.

Hamilton, Lawrence. "Concern about Toxic Wastes: Three Demographic Predictors." Paper presented at the American Sociological Association, August, 1985a.

_____. "Who Cares about Pollution: Opinions in a Small-Town Crisis." *Sociological Inquiry,* May(1985b).

Hardin, Garrett. "The Tragedy of the Commons." *Science,* 162(1968), 1243–1248.

Harmon, Willis. *An Incomplete Guide to the Future.* San Francisco: San Francisco Book Co., 1976.

Harris, David. "Health Department: Enemy or Champion of the People?" *American Journal of Public Health,* 74(1984), 428–430.

Hatcher, Sherry Lynn. "The Psychological Experience of Nursing Mothers upon Learning of a Toxic Substance in Their Breast Milk." *Psychiatry,* 45(1982), 172–181.

Hayes, Pat. Comments in A. Armour (Ed.), *The Not-in-my-backyard Syndrome.* Downsview, Ontario: York University, 1984, 15–17.

Hayword, D. Geoffrey. "An Overview of Psychological Concepts of 'Home'." Paper presented at the Environmental Design Research Association Conference, Champaign-Urbana, Ill., April, 1977.

————. "Home as an Environmental and Psychological Concept." *Landscape,* 20(1976), 2–9.

Heider, Fritz. *The Psychology of Interpersonal Relations.* New York: John Wiley and Sons, 1958.

Heller, Kenneth, Richard Price, Shulamit Reinharz, Stephanie Riger and Abraham Wandersman. *Psychology and Community Change: Challenges of the Future.* Homewood, Ill.: Dorsey Press, 1984.

Hewitt, Richard. "Toxic Waste: A Long, Winding Freight Train to Nowhere." *Times Herald Record* (Middletown, New York), June 28, 1981.

Hill, Gladwin. "Stringfellow Toxins May Be Headed for Tap Water Supply: Big Test Ahead for E.P.A. Superfund." *The New York Times,* Sunday, August 26, 1984.

Hinchey, Maurice. "Organized Crime's Involvement in the Waste Hauling Industry." Report to the New York State Assembly Environmetal Conservation Committee, July 25, 1986.

Hohenemser, C., R. W. Kates and P. Slovic. "The Nature of Technological Hazard." *Science,* April 22 (1983), 376–383.

Hollander, E. P. "Conformity, Status, and Idiosyncracy Credit." *Psychological Review,* 65(1958), 117–127.

Holman, Thomas. "The Influence of Community Involvement on Marital Quality." *Journal of Marriage and the Family,* February (1981), 143–149.

Ibsen, Henrik. *An Enemy of the People.* Adapted by Arthur Miller. New York: Penguin Books, 1979.

ICF, Inc. "Analysis of Community Involvement in Hazardous Waste Site Problems." A report to the Office of Emergency and Remedial Response, U.S. Environmental Protection Agency, July, 1981.

Illich, Ivan. "Disabling Professions" in Ivan Illich, Irving Zola, John McKnight, Jonathan Caplan and Harley Shaiken, *Disabling Professions.* London: Marion Boyars, 1977.

Isaacs, Colin. Comments in A. Armour (Ed.), *The Not-in-my-backyard Syndrome.* Downsview, Ontario: York University, 1984, 7.

Janis, Irving. *Stress and Frustration.* New York: Harcourt, Brace and Jovanovich, 1971.

Janis, Irving and Leon Mann. *Decision Making.* N.Y.: The Free Press, 1977.

Janoff-Bulman, Ronnie and Irene Hanson Frieze. "A Theoretical Perspective for Understanding Reactions to Victimization." *Journal of Social Issues,* 39, No. 2(1983), 1–17.

Jones, Edward, Amerigo Farina, Albert Hastorf, Hazel Markus, Dale Miller and Robert Scott. *Social Stigma: The Psychology of Marked Relationships.* New York: W. H. Freeman, 1984.

Jordan, W. S., Jr. "Psychological Harm Done After PANE: NEPA's Requirement to Consider Psychological Damage." *The Harvard Law Review,* 8(1984), 1.

Kameron, Joel. *Man Nature Value Orientations.* Dissertation from the City University of New York, 1975.

Kasperson, Roger (Ed.). *Equity Issues in Radioactive Waste Management.* Cambridge, Mass.: Oelgeschlager Gunn, & Hain, 1983.

Katz, Daniel and Robert Kahn. *The Social Psychology of Organizations,* 2d edition. New York: John Wiley and Sons, 1978.

Kelley, Harold. "Attribution in Social Interaction" in E. Jones, D. Kanouse, H. Kelley, R. Nisbett, S. Valins and B. Weiner, *Attribution: Perceiving the Causes of Behavior.* Morristown, N.J.: General Learning Press, 1972, 197.

Kim, Nancy and Daniel Stone. "Organic Chemicals and Drinking Water." Report to the New York State Department of Health, 1980.

Klausner, S. L. *On Man and His Environment.* San Francisco: Jossey-Bass, 1971.

Kleese, Deborah. "Mother as a Mediator of Environmental Hazards."*Childhood City Quarterly,* 9, No. 3(1982), 3–7.

Koffka, Kurt. *Principles of Gestalt Psychology.* New York: Harcourt Brace, 1935.

Krawetz, Natalia. "Hazardous Waste Management: A Review of Social Concerns and Aspects of Public Involvement." Staff Report 4, Alberta Environment, November, 1979.

Kroll-Smith, J. Stephen and Samuel Garula, Jr. "The Real Disaster Is Above Ground: Community Conflict and Grass Roots Organization in Centralia." *Small Town,* 15(1985), 4–11.

Kroll-Smith, J. Stephen and Stephen Couch. "Fear and Suspicion in Centralia: Doing Fieldwork in a Community in Crisis." Paper presented at the Society for the Study of Social Problems Conference, San Antonio, Texas, August, 1984.

Kuhn, Thomas. *The Structure of Scientific Revolutions.* Chicago: University of Chicago Press, 1962.

Kulik, James and Heike Mahler. "Health Status, Perceptions of Risk, and Prevention Interest for Health and Nonhealth Problems." *Health Psychology,* 6, No. 1(1987), 15–27.

Kushnir, Talma. "Skylab Effects: Psychological Reactions to a Human-made Environmental Hazard." *Environment and Behavior,* 14, No. 1(1982), 84–93.

Landy, Mark. "Cleaning Up Superfund." *The Public Interest,* 85(1986), 58–71.

Lang, Kurt and G. E. Lang. "Collective Responses to the Threats of Disaster" in G. H. Grosser, H. Wechsler and M. Greenblatt (Eds.), *The Threat of Impending Disaster.* Cambridge, Mass.: The M.I.T. Press, 1964, 58–75.

Lasch, Jonathan, et al. *A Season of Spoils: The Story of the Reagan Administration's Attack on the Environment.* New York: Pantheon, 1984.

Lazarus, Richard. "The Study of Psychological Stress: A Summary of Theoretical and Experimental Findings" in Charles Speilberger (Ed.), *Anxiety and Behavior.* New York: Academic Press, 1966, 225–231.

———. "A Laboratory Approach to Psychological Stress" in G. H. Grosser, H. Wechsler and M. Greenblatt (Eds.), *The Threat of Impending Disaster.* Cambridge, Mass.: The M.I.T. Press, 1964, 34–57.

Lazarus, Richard and R. Launier. "Stress-related Transactions between Person and Environment" in L. A. Pervin and M. Lewis (Eds.), *Perspectives in Interactional Psychology.* New York: Plenum Publishers, 1978, 287–327.

Lee, Brenda. "The Social Impact Assessment of Hazardous Waste Management Facilities: Covering the Bases." Publication of the University of Toronto, March, 1984.

Lee, Terrence. "The Perception of Risks" in The Royal Society, *Risk Assessment: A Study Group Report*, Inprint of Luton Limited, Luton, Belfordshire, U.K., 1983.

Lerner, Melvin J. *The Belief in a Just World: A Fundamental Delusion.* New York: Plenum Press, 1980.

Levine, Adeline. *Love Canal: Science, Politics and People.* Boston, Mass.: Lexington Books, 1982.

Lindell, Michael, Timothy Earle and S. M. Nealey. "Comparative Analysis of Risk Characteristics of Nuclear Waste Repositories and Other Disposal Facilities." Battelle Human Affairs Research Center report, June, 1981.

Lumsden, D. Paul. "Towards a Systems Model of Stress: Anthropological Study of the Impact of Ghana's Volta River Project" in I. Sarason and C. Spielberger (Eds.), *Stress and Anxiety, Volume 2.* New York: John Wiley and Sons, 1975, 191–228.

Macdonald, H. Ian. Welcome address in A. Armour (Ed.), *The Not-in-my-backyard Syndrome.* Downsview, Ontario: York University Press, 1984.

Madisso, Urmas. "A Synthesis of Social and Psychological Effects of Exposure to Hazardous Substances." Report to the Inland Waters Directorate, Ontario Region, January, 1985.

———. "An Annotated Bibliography of the Literature on the Social and Psychological Effects of Exposure to Hazardous Substances." Report to the Inland Waters Directorate, Ontario Region, September, 1984.

Marshall, Eliot. "EPA's High Risk Carcinogen Policy." *Science*, 218(Dec., 1982), 975–978.

Marx, Leo. *The Machine in the Garden: Technology and the Pastoral Ideal in America.* New York: Oxford University Press, 1964.

Mazur, Alan. *The Dynamics of Technical Controversy.* Washington, D.C.: Communications Press, 1981.

Melief, Willem. "The Social Impacts of Alternative Policy Approaches to Incidents of Toxic Waste Exposure" in Henk Becker and Alan Porter (Eds.), *Impact Assessment Today, Vol. 11.* Utrecht, The Netherlands: Jan van Arkel, 1986, 825–834.

Metropolitan Edison Company, et al. and United States Nuclear Regulatory Commission, et al., v. People Against Nuclear Energy, et al., Supreme Court opinion by Justice Rehnquist, *The United States Law Week,* 4-19-83, 51 LW 4371–4375.

Milbrath, Lester. *Environmentalists: Vanguard for a New Society.* Albany: State University of New York Press, 1984.

Miller, David T., Sr. "One View from Jackson." *The Jackson News (Jackson Township, New Jersey),* Tuesday, April 28, 1981.

Miller, J. G. "A Theoretical Review of Individual and Group Psychological Reactions to Stress" in G. H. Grosser, H. Wechsler and M. Greenblatt (Eds.), *The Threat of Impending Disaster.* Cambridge, Mass.: The M.I.T. Press, 1964, 11–33.

Miller, Robert. "I'm from the Government and I'm Here to Help You: Fieldwork at Times Beach and Other Missouri Dioxin Sites." Paper delivered at the Society for the Study of Social Problems, San Antonio, Texas, August, 1984.

Molotch, Harvey and Marilyn Lester. "Accidental News: The Great Oil Spill as Local Occurrence and National Event." *American Journal of Sociology*, 81(1975), 235–260.

Monat, Alan and Richard Lazarus (Eds.). *Stress and Coping: An Anthology.* New York: Columbia University Press, 1977.

Montague, Peter. "The Limitations of Landfilling" in Bruce Piasecki (Ed.), *Beyond Dumping: New Strategies for Controlling Toxic Contamination.* Westport, Conn.: Greenwood Press, 1984.

Morrison, Denton. "Doomseers, Boomseers, Techseers, Socseers: The Future as Seen in the Rearview Mirror." *World Future Society Bulletin*, Nov.-Dec., (1983), 7–13.

Morrison, Denton and Riley Dunlap. "Is Environmentalism Elitist?" in Frederick Buttel and Craig Humphrey (Eds.), *Environment and Society.* University Park: Pennsylvania State University Press, 1985.

Mueller, Claus. *The Politics of Communication.* New York: Oxford University Press, 1973.

Norris, Ruth (Ed.). *Pills, Pesticides and Profits: International Trade and Toxic Substances.* Croton-on-Hudson, New York: North River Press, 1982.

Novick, Sheldon. "What is Wrong with Superfund?" *The Environmental Forum*, 2(1983), 6–11.

Ocean County Department of Planning. "General Statistical Information, Ocean County, New Jersey," 1981.

Office of Technology Assessment. *Technologies and Management Strategies for Hazardous Waste Control.* Washington, D.C.: Congress of the United States, March, 1983.

Ottum, Margaret and Nancy Updegraff. "Local Residents' Perception of the Williamstown Pollution Problem." Unpublished research summary, 1984.

Paigen, Beverly. "The Ethical Dimensions of Scientific Conflict: Controversy at Love Canal." *Hastings Center Report*, June (1982), 29–37.

———. "Health Hazards at Love Canal." Testimony presented to the House Subcommittee on Oversight and Investigations, March 21, 1979.

Pearlin, Leonard, Elizabeth Menaghan, Morton Lieberman and Joseph Mullan. "The Stress Process." *Journal of Health and Social Behavior*, 22(1981), 337–356.

Perin, Constance. *Everything in its Place: Social Order and Land Use in America.* Princeton, N.J.: Princeton University Press, 1977.

Perry, R. W., H. Parker and D. Gillespie. *Social Movements and the Local Community.* Beverly Hills, Ca.: Sage Publications, 1976.

Peterson, Christopher and Martin Seligman. "Learned Helplessness and Victimization." *Journal of Social Issues*, 39, No. 2(1983), 103–116.

Pirages, Dennis. *The New Context for International Relations: Global Ecopolitics.* North Scituate, Mass.: Duxbury Press, 1978.

Polanyi, Karl. *The Great Transformation.* Boston: Beacon Press, 1944.

Preston, Valerie, S. Martin Taylor and David Hodge. "Adjustment to Natural and Technological Hazards: A Study of an Urban Community." *Environment and Behavior,* 15(1983), 143–164.

Purcell, Arthur. "Setting Priorities in Managing Toxic and Hazardous Substances." *The Environmental Professional,* 3(1981), 9–13.

Quarantelli, E. L. and Russell Dynes. "Response to Social Crisis and Disaster." *Annual Review of Sociology,* 3(1977), 23–49.

Rapoport, Amos. *House Form and Culture.* Englewood Cliffs, N.J.: Prentice-Hall, 1969.

Reich, Michael. "Environmental Politics and Science: The Case of PBB Contamination in Michigan." *American Journal of Public Health,* 73, No. 3(1983), 302–313.

Reim, B., D. Glass and J. Singer. "Behavioral Consequences of Exposure to Uncontrollable and Unpredictable Noise." *Journal of Applied Social Psychology,* 1(1971), 44.

Reko, H. Karl. "Not an Act of God: The Story of Times Beach." Privately published report, 1984.

Ridley, Scott. *The State of the States.* Washington, D.C.: Fund for Renewable Energy and Environment, 1987.

Robertson, John. "Geohydrologic Aspects of Hazardous Waste Disposal" in Denise Wiltshire and Dan Hahl (Eds.), *Information Needs for Tomorrow's Priority Water Issues.* Albany: U.S. Geological Survey, 1983.

Rogers, Ronald. "A Protection Motivation Theory of Fear Appeals and Attitude Change." *The Journal of Psychology,* 91(1975), 93–114.

Ruesch, Jurgen and Weldon Kees. *Nonverbal Communication.* Berkeley: University of California Press, 1956.

Ryan, W. *Blaming the Victim.* New York: Pantheon, 1971.

Sarason, Seymour. *The Psychological Sense of Community.* Washington, D.C.: Jossey-Bass, 1974.

Schacter, Sidney. *The Psychology of Affiliation.* Stanford, Ca.: Stanford University Press, 1959.

Scott, W. and M. Wertheimer. *Introduction to Psychological Research.* New York: John Wiley and Sons, 1962.

Seley, John and Julian Wolpert. "Equity and Location" in Roger E. Kasperson (Ed.), *Equity Issues in Radioactive Waste Management.* Cambridge, Mass.: Oelgeschlager Gunn, & Hain, 1983, 69–93.

Selye, Hans. *The Stress of Life.* New York: McGraw Hill, 1976.

Shaw, L. G. and Lester W. Milbrath. "Citizen Participation in Governmental Decision Making: The Toxic Waste Threat at Love Canal, Niagara Falls, New York." *Rockefeller Institute Working Papers,* No. 8(1983).

Shrivastava, Paul. *Bhopal: Anatomy of a Crisis.* Cambridge, Mass.: Ballinger Publishing Co., 1987.

Slater, Philip. *Earthwalk.* Garden City, N.Y.: Doubleday, 1974.

Slovic, Paul, Baruch Frischhoff and Sarah Lichtenstein. "Facts and Fears: Understanding Perceived Risk" in R. Schwing and W. A. Albers, Jr. (Eds.), *Societal Risk Assessment: How Safe Is Safe Enough?* New York: Plenum Press, 1980, 181–214.

Sonnenfeld, J. "Values in Space and Landscape." *Journal of Social Issues*, October (1966), 71.

Sorenson, John, Jon Soderstrom, Emily Copenhaver, Sam Carnes, and Robert Bolin. *Impacts of Hazardous Technology: The Psycho-Social Effects of Restarting TMI-1.* Albany: State University of New York Press, 1987.

Sowder, Barbara (Ed.). *Disasters and Mental Health: Selected Contemporary Perspectives.* Rockville, Md.: National Institute of Mental Health, 1985.

Stallen, Pieter Jan and Arend Tomas. "Public Concern about Industrial Hazards." Paper presented at the Annual Meeting of the Society for Risk Analysis, Washington, D.C., October, 1985.

Stone, Russell and Adeline Levine. "Reactions to Collective Stress: Correlates of Active Citizen Participation at Love Canal" in Abraham Wandersman and Robert Hess (Eds.), *Beyond the Individual: Environmental Approaches and Prevention.* New York: Haworth, 1985, 153–178.

Stone, Russell and Adeline Levine. "Residents' Perceptions of the Hazard of Love Canal: Problems and Changes to Self and Family." Paper presented at the American Psychological Association, Toronto, Canada, August, 1984.

Sunday Star Ledger (Newark, New Jersey) March 11, 1984.

Tester, David. Panel presentation at the session on impacts of regulatory proceedings, International Association for Impact Assessment, Vancouver, B.C., October, 1982.

Times Herald Record (Middletown, New York), February 25, 1985.

Times Herald Record (Middletown, New York), February 6, 1984.

Timmerman, Peter. "Ethics and Hazardous Waste Facility Siting." Publication of the University of Toronto, March, 1984a.

———. "Ethics and the Problem of Hazardous Waste Management: An Inquiry into Methods and Approaches." Publication of the University of Toronto, March, 1984b.

Toth, Robert. "Life Without Chemicals: Does Bad Outweigh Good?" in *Training Manual on Toxic Substances, Book One.* San Francisco: Sierra Club, 1981.

Unger, David and Abraham Wandersman. "The Importance of Neighboring: The Social, Cognitive, and Affective Components of Neighboring." *American Journal of Community Psychology*, 13(1985), 139–169.

van Eijndhoven, J.C.M. and G.H.E. Nieuwdorp. "Institutional Action in Soil Pollution Situations with Uncertain Risks" in Henk Becker and Alan Porter (Eds.), *Impact Assessment Today, Vol. II.* Utrecht, The Netherlands: Jan van Arkel, 1986, 267–278.

van Eijndhoven, J.C.M., D. Hortensius, C. Nauta, G.H.E. Nieuwdorp and C.W. Worrell. "Hazardous Waste in the Netherlands: Dutch Policies from a Local Perspective." Report to the International Institute for Applied Systems Analysis, March, 1985.

Vissing, Yvonne. "The Difficulties in Determining Elite Deviance: Dow Chemical Company and the Dioxin Controversy." Paper presented at the Society for the Study of Social Problems, San Antonio, Texas, August, 1984.

von Uexküll, Thure. "Ambient and Environment—or Which is the Correct Perspective on Nature?" Paper presented at the International Association for the Study of People and Their Physical Surroundings, West Berlin, West Germany, July 25–29, 1984.

Vyner, Henry. "The Psychological Effects of Invisible Environmental Contaminants." *Social Impact Assessment,* (July-September, 1984), 93–95.

Wallace, Anthony. "Mazeway Disintegration." *Human Organization,* 14(1957), 23–27.

Walsh, E. J. and R. H. Warland. "Social Movement Involvement in the Wake of a Nuclear Accident." *American Sociological Review,* 48(1983), 764–780.

Wandersman, Abraham. "A Framework of Participation in Community Organizations." *Journal of Applied Behavioral Science,* 17(1981), 27–58.

Weinberger, Morris, James Greene, Joseph Mamlin and Michael Jerin. "Health Beliefs and Smoking Behavior." *American Journal of Public Health,* 71, No. 11(1981), 1253–1255.

Weinstein, Neil. "Unrealistic Optimism About Susceptibility to Health Problems." *Journal of Behavioral Medicine,* 5, No. 4(1982), 441–460.

Weller, Phil. Comments in A. Armour (ed.), *The Not-in-my-backyard Syndrome.* Downsview, Ontario: York University, 1984, 76–78.

White, Lynn, Jr. "The Historical Roots of Our Ecologic Crisis." *Science,* 155(1967), 1203–1207.

Wicker, Allan. *An Introduction to Ecological Psychology.* Monterey, Ca.: Brooks Cole Publishing Co., 1979.

Wilkinson, Charles. "Aftermath of a Disaster: The Collapse of the Hyatt Regency Hotel Skywalks." *American Journal of Psychiatry,* 140(1983), 1134–1139.

Wolfenstein, Martha. *Disaster: A Psychological Essay.* Glencoe, Ill.: Free Press, 1957.

Wohlwill, Joachim. "The Physical Environment: A Problem for a Psychology of Stimulation." *Journal of Social Issues,* 22, No. 4(1966), 29.

Wohlwill, Joachim and Imre Kohn. "The Environment as Experienced by the Migrant: An Adaptation-Level View." *Representative Research in Social Psychology,* 4(1973), 135–164.

Index

266224BV00002B/13/P
Printed in the USA
CPSIA information can be obtained at www.ICGtesting.com